JUICE

JUICE

A HISTORY OF FEMALE EJACULATION

STEPHANIE HAERDLE
TRANSLATED BY ELISABETH LAUFFER

The MIT Press
Cambridge, Massachusetts
London, England

Originally published as *Spritzen: Geschichte der weiblichen Ejakulation* © Edition Nautilus GmbH 2019

This book was set in Bembo by Jen Jackowitz. Printed and bound in the United States of America.

Library of Congress Cataloging-in-Publication Data

Names: Haerdle, Stephanie, 1974- author. | Lauffer, Elisabeth, translator.
Title: Juice : a history of female ejaculation / Stephanie Haerdle ; translated by Elisabeth Lauffer.
Description: Cambridge, Massachusetts : The MIT Press, [2024] | "Originally published as Spritzen: Geschichte der weiblichen Ejakulation, Edition Nautilus GmbH 2019." | Includes bibliographical references.
Identifiers: LCCN 2023032400 | ISBN 9780262048514 (paperback) | ISBN 9780262381697 (epub) | ISBN 9780262381680 (pdf)
Subjects: LCSH: Women—History. | Ejaculation.
Classification: LCC HQ1121 .H3413 2024 | DDC 305.409—dc23/eng/20230714
LC record available at https://lccn.loc.gov/2023032400

10 9 8 7 6 5 4 3 2 1

If only I had courage enuf to kill myself when you reach the climax then—then I would have known happiness, for then at that moment I had complete possession of you. . . .

I cannot escape from the rhythmic spurt of your love juice. . . . my dear whose succulence is sweet and who dreeps with honey.
—Love letter from Almeda Sperry to anarchist
Emma Goldman

CONTENTS

AUTHOR'S NOTE

This book traces the cultural history of what I want to refer to as *juice*: that surge of fluid at the height of sexual pleasure in some women and people with vulvas. Telling this story hasn't been easy, as attitudes toward these effusions and the language used to describe them have forever been in flux. Although lovers had delighted in women's wetness for thousands of years, it later came to be known as "seed" or "pollution," diluted and degraded, reduced to its role in conception and fertility.

For more than a century, the Western world used the term "female ejaculation" for this physical phenomenon and its attendant efflux, but in the last ten years or so, there's been a major shift in our understanding. Recent scientific studies have concluded that during sex, people with vulvas may spurt two different fluids produced in two different organs. In short, we're looking at two different things: female ejaculation and squirting.

I believe in manifold genders. The separation of humans into just two categories—man and woman—with mutually exclusive skills, traits, and behaviors is problematic and scientifically obsolete. The concept of a gender binary docks identities, hinders diversity, and creates or upholds hierarchies. In exploring the history of genital sexual fluids produced by individuals seen as women, I am compelled to work with these very categories, as they have been employed for millennia to view, comprehend, and explain the world. I use the term female ejaculation as it is one of the two bodily functions scientific studies have identified, although I recognize the error of linking anatomy to gender. What this term does, though, is call male ejaculation to mind. And on making this connection, we wind up caught between visibility and invisibility, known and unknown, swarmed by questions.

For a long time, the patriarchal conception of bodies has employed language to obscure certain things by *not* naming them. "Female ejaculation" and the "female prostate" allude directly to their male equivalents, which highlights similarities between human bodies. On the other hand, they also perpetuate the notion of a gender binary, and attach masculinity and femininity to anatomy. In an effort to disrupt this pattern, I also use the term "vulval ejaculation."

In the studies cited in this book, test subjects were cis women (and men). As far as I know, when I wrote *Juice* in 2018, there had not been any research done on ejaculation and squirting that took trans people into consideration. As recently as 2021, a systematic review of sexual health in people after vaginoplasty included just one brief paragraph on the topics of self-lubrication, secretions, and ejaculation.[1]

In a world where the openness and fluidity of sex and gender are familiar and widely accepted notions, all we'll have to say is "ejaculation" and "prostate." Everyone will know that what we're talking about is common to folks of all genders.

PREFACE

> Is it so frightening to believe that woman can, in a sense, ejaculate too?
>
> —Juliet Richters, "Bodies, Pleasure and Displeasure"

> Society cannot accept female ejaculation precisely because it makes men and women equal.
>
> —Fanny Fatale, quoted in Rebecca Chalker, *The Clitoral Truth*

Wait, women can ejaculate?

That's right! Up to 69 percent of those born biologically female release fluid on climaxing.[1] Whether someone with a vulva spurts a scant teaspoon of liquid or the sheets need wringing out after the act, lots of folks love this aspect of female sexuality. According to a 2013 study, 78.8 percent of women and 90 percent of their partners describe their ejaculation as "enriching their sex life."[2] Nevertheless, female ejaculation remains a controversial topic, its very existence disputed by some. What skeptics consider the stuff of myth, meanwhile, is just everyday sex to others.

So what do we really know about this aspect of female pleasure? What does the research show, and why are so many details still unknown? Does a history of female ejaculation exist? What did other eras know or think about the flow or gush of sexual fluids of persons with vulva? What knowledge was forgotten, and why? How has the phenomenon been interpreted and (mis)used? How did earlier cultures frame these effusions? For instance, where did vulval juices fit into depictions of bodies, arousal, sex, and procreation in Chinese or Indian erotica of the past? Were classical civilizations aware of women's satiated surge, and what did the (almost exclusively male) doctors, philosophers, and writers of the Middle Ages and early modern period make of it? How were female emissions *rediscovered*, and how, and what beliefs, fantasies, and fears accompanied this reappraisal?

What does a cultural history of expressive genital female juices look like?

The search for accounts of female juices extends around the globe and far into the pre-Christian past. The findings are surprising: for thousands of years, it went without saying that both women and men ejaculated during intercourse. In the late nineteenth century, however, female ejaculation came under fire in Europe, the fact of its existence denied, contested, suppressed, recast as taboo, and ultimately forgotten by most.

Exploring past attitudes and cultures that considered ejaculation and gushing a natural part of female sexuality is just one compelling aspect of this history.[3] It's also fascinating to learn why vulval ejaculation has routinely been overlooked, dismissed, or banished to the realm of the imaginary as just another "male sex fantasy," to the point where the very idea of it seems obscene.[4]

The history of female gushes is also a history of women, their desires, and the worship and denigration of the female body. Many cultures believed a woman's sexual effluence corresponded to male ejaculate. At times these respective fluids

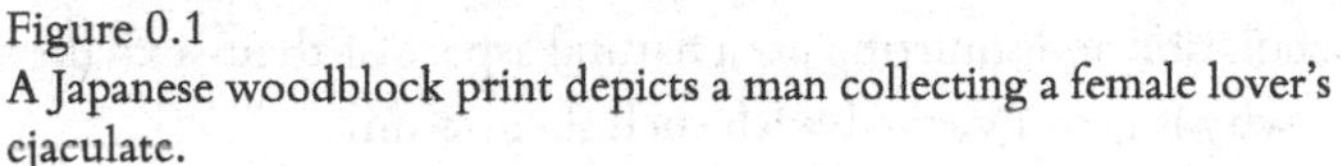

Figure 0.1
A Japanese woodblock print depicts a man collecting a female lover's ejaculate.

were valued equally, while at others, one might be elevated over the other; either way, they were considered complementary "generative matter" in the act of conception. In cultures that saw similarities between female and male bodies, and placed great value on sex and female desire, women were known to express liquid at orgasm. When sperm and egg cells first became visible under the microscope, and humans gained an understanding of the reproductive process, female ejaculation vanished—if not from bedrooms, then from medical discourse, which dictated the understanding of this fluid. Once the ovum was identified as women's contribution to baby y-making, "female semen" was rendered meaningless. As Michel Foucault notes, "Nothing that was not ordered in terms of generation or transfigured by it could expect sanction or protection. Nor did it merit a hearing. It would be driven out, denied, and reduced to silence. Not only did it not exist, it had no right to exist . . . ."[5] Doctors in the nineteenth and early twentieth centuries, like English physician William Acton or Italian surgeon-turned-criminologist Cesare Lombroso, speculated whether women even experienced sexual urges. The suppression of female desire was one of the first steps toward shrouding female ejaculation in uncertainty. Then sex became taboo, female and male bodies were cast as polar opposites, the "dogma of the complementary sex" gained purchase, and hostility toward the vagina itself emerged among some contingents of second-wave feminism.[6] In the face of such forces, female ejaculation was transformed into outright myth.

For countless women and those who have vulvas, however, ejaculation and squirting *are* a natural aspect of their sexuality. So why is it still viewed with such skepticism?

ONE STEP FORWARD, TWO STEPS BACK

Numerous studies of female ejaculation have been conducted since the 1980s. Medical researchers have looked into the anatomical or biochemical aspects of the phenomenon; they've performed endoscopies and radiology exams, ultrasounds, and MRIs to get to the bottom of it. Nevertheless, the jury is out among these many sexologists, urologists, pathologists, anatomists, and gynecologists. To this day, there are conflicting explanations for exactly where and how the female body produces this fluid, and exactly where and how women expel it. Vexingly, new findings on ejaculation and the parts of female anatomy most closely tied to it—the female prostate, clitoral complex, and urethra—are routinely "forgotten" in the literature. In 2001, for instance, the Federative International Committee for Anatomical Terminology (FICAT), whose mission was to establish a standardized global term base for medicine, updated its nomenclature guide with the term "female prostate."[7] In an about-face almost twenty years later, the Federative International Programme for Anatomical Terminology (FIPAT), a successor of FICAT, adopted the term "para-urethral glands of female urethra." "Female prostate," meanwhile, was relegated to the column "Other," alongside "Skene's glands" and "prostata feminina."[8]

If you consult current medical guides, textbooks, or popular online platforms for information on the female prostate, prepare for disappointment: if it comes up at all, these sources employ facile, nonstandardized terms for the anatomical feature. Rarely do they mention the fact that the female prostate is a fully functioning gland, homologous—that is, based on the same embryological system—with the male prostate.[9] By and large, the prostate and ejaculation are discussed exclusively as they relate to the male body.[10] Wikipedia and WebMD articles on the prostate only cover the male gland.

Medicine and anatomy have never been stable sciences; rather, these fields are influenced by social, cultural, political, and economic factors. Long the purview of men, medicine and anatomy were shaped by male perspectives, wishes, and needs. Structures of the female body at odds with specific notions of femininity were dismissed or ignored. Societal contempt for women was reflected in a lack of interest in the female body and its anatomy, sexual responses, and desires. Women's bodies were long viewed as lesser versions of men's bodies. Faint echoes of these attitudes can still be detected in foundational medical texts used today. The female urethra is "only" one to two inches long. The vaginal wall consists of "weak" muscles.[11] The clitoris is the "evolutionary counterpart to the penis."[12] (US psychiatrist Mary Jane Sherfey, meanwhile, argues that male genitalia develop from an "embryonic female morphology" and the penis can therefore be said to represent an "exaggerated clitoris.")[13]

With women mostly barred from the sciences at the time, perhaps these male scientists saw what they wanted to see, and personal areas of interest were financed and researched. Anna Fischer-Dückelmann, one of the first women in German-speaking countries to study medicine, lamented the status quo in her 1900 bestseller, *Das Geschlechtsleben des Weibes* (The sexual life of women): "Scientifically sound studies of sexual life thus far have been conducted on men alone, for the use of men alone. Men alone have studied sexual life; men alone have made women the object of study."[14] For a surprisingly long time, female sexual anatomy and desire have received little attention. Even today, these topics are considered an "obscure and unknown part of medicine."[15] Female genitalia have been defined and described largely with regard to their role in reproduction and the sexual function they perform for men (read: vaginal penetration). As one medical publication from 1800 explains, "The entrance to the vagina is sized appropriately for the male member."[16]

Depictions of female genitalia in anatomy books were left incomplete for centuries. Allowing these representations to disappear from medical discourse and public awareness exemplifies the androcentric "study" of the female body; sex educator and author Rebecca Chalker has described this disappearance as "one of the grand heists of all time."[17] Over the course of the twentieth century, we learned that the clitoral glans (or crown) is just the visible part of a much larger, more complex organ that swells when aroused. In Alfred Benninghoff and Kurt Goerttler's 1957 textbook *Anatomie des Menschen*

(Human anatomy), the authors provided a precise rendering and description of the human clitoris, which would be reproduced in *Sexuologie, Geschlecht, Mensch, Gesellschaft* (Sexology: Gender, humans, and society), a three-part reference work published in the German Democratic Republic in 1974.[18] In the 1960s, Sherfey described the structures of the clitoris that extend deep into the body. As part of the women's health movement in the 1980s, a new representation of the clitoris was disseminated in such best-selling works as *A New View of a Woman's Body*, put out by the American Federation of Feminist Women's Health Centers. In the late 1990s, Australian urologist Helen O'Connell sparked a media blitz with her "revolutionary discovery" of the inner clitoris. Nevertheless, most people still believe that the visible bit of the clitoris is *the* clitoris, while the actual organ is more than 2 inches wide and extends up to 3.5 inches into the pelvic area. In addition to the crown and hood, it includes the body, two leglike crura, and a pair of vestibular bulbs, whose spongy erectile tissue swells when the woman is aroused. The crown exhibits "the highest-known number of sensory nerve endings."[19] And it's only the tip of the iceberg.

Sylvia Groth and Kerstin Pirker, who run a women's health center in Graz, Austria, sum it up in a 2009 article in the women's health journal *clio*:

> These experiences and the knowledge of the clitoris, vulva, and vagina gained by means of self-examination during the women's health movement of the '70s did not make their way into sex education resources, nor can they be found in news media, schoolbooks, literature, or film. This new understanding of the clitoris and female sexuality has barely been incorporated into training for educators and health care professionals, like doctors.[20]

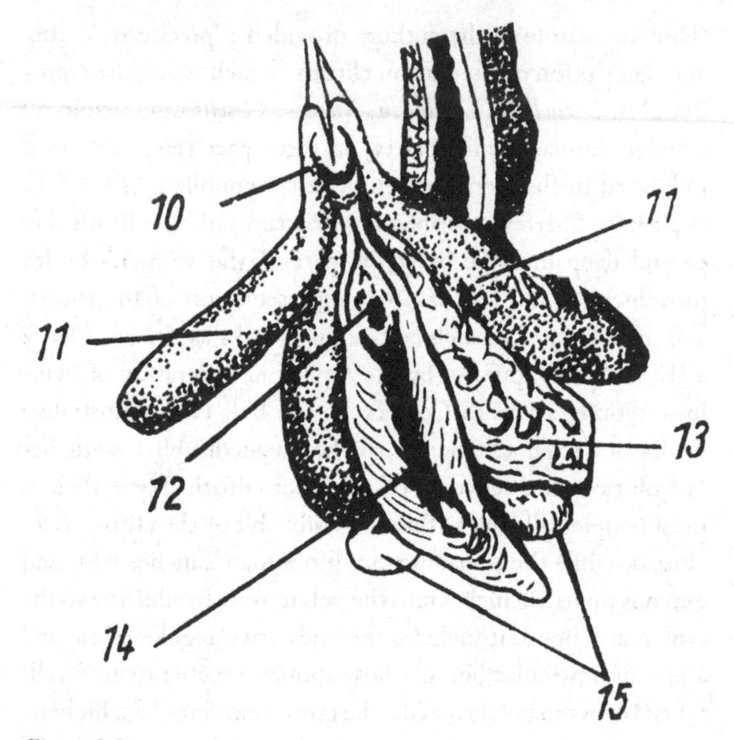

Figure 0.2
Diagram from *Sexuologie: Geschlecht, Mensch, Gesellschaft* (Sexology: Gender, humans, and society) (1974): 10 *glans clitoridis* / glans of clitoris; 11 *crus clitoridis* / erectile crus of clitoris; 12 urethral opening; 13 vestibular bulbs; 14 vaginal opening; and 15 Bartholin's glands.

Unfortunately, one of the only places such knowledge *is* shared freely and without shame—a place where sexual and body diversity is celebrated, and Peaches's hit "Fuck the Pain Away" is the school song—is the fictional Moordale Secondary, from the English coming-of-age series *Sex Education*. At Moordale, bake sales feature colorful vulva cupcakes, while at one assembly, students declare their sexual idiosyncrasies with handmade signs around their necks, including one that reads "I am a squirter," complete with cartoon water droplets.[21]

Communicating science news to the public routinely falters, but often experts themselves prove underinformed of significant research findings. Viennese urologist Florian Wimpissinger, who studies and writes about the female prostate and ejaculation, comments, "It's interesting that specialists versed in anatomy and surgery—experts in the fields of urology and gynecology as well as anatomy—are usually unable to confirm the existence of the female prostate with any degree of certainty."[22] Or to quote Nancy Fish, a US specialist in vulvodynia, or vaginal pain, "We are *so* in the Dark Ages when it comes to medical care and understanding in the area of the vagina."[23] Who'd have thought that the anatomy and physiology of the vagina, clitoris, and urethra would remain partially unknown into the twenty-first century? Disparate takes on the female prostate and ejaculation reflect this patchy knowledge and paltry or contradictory understanding of women's sexual physiology.

Another stumbling block in the study of vulval ejaculation is the medical separation of female urinary and reproductive systems. The two have been studied and described as "utterly discrete *in function*," meaning urologists are responsible for one set of organs and the surrounding structures, and gynecologists for the other.[24] And this is despite the fact that these internal systems actually both emerge from the urogenital sinus—a structure present in the human embryo during development of the urinary and reproductive organs—which means the uterus, vagina, and urethra are in fact closely connected. The entire length of the urethra, for instance, is embedded in the connective tissue of the anterior vaginal wall. To understand vulval ejaculation and squirting, urological and gynecological research must be considered in concert, and the clitoral complex, urethra, prostate, and vagina must be regarded as a unit, both anatomically and functionally.[25] Once this framework has been established, it becomes clear that the female urethra is a sex organ, as psychologist Josephine Lowndes Sevely and others have demonstrated.

IN THE BEGINNING WAS THE WORD

We use names to identify and exchange ideas as well as "sharpen the awareness of one part as distinct from another."[26] Ejaculation derives from the Latin *eiaculari*, meaning "to throw out" or "expel," and refers to the emission of fluid that often—but not always—occurs during orgasm. Although sperm accounts for less than 1 percent of the total volume of male ejaculate, and "ejaculation" is also used when the fluid contains no sperm at all (such as that produced in prepubescent boys), the term is generally understood to mean the emission of semen, itself colloquially synonymous with sperm. And what about individuals who have had vasectomies? Don't they still ejaculate? No one would ever "dream of denying ejaculation to a man who's had his vas deferens cut."[27] If men's and women's orgasms are nearly identical—or homologous—at the physiological level, why shouldn't both be referred to as ejaculation? If the term "ejaculation" is used to signify the "discharge of sexual fluids," then it really could be used for women and men alike.

When combing historical texts for insights into female juices, one quickly comes up against the issue of translation. Translators have rarely proven familiar with the act now known as ejaculation, and squirting and the fluids, which are frequently lost in the rendering. At the hands of these penmen, a woman's excited spray has been turned into "vaginal fluid," "vaginal lubrication," "effluence," "gonorrhea," "leukorrhea," "mucus," or "juice." And "over time, that which goes unnamed ceases to be seen or experienced," Sabine zur Nieden argues in her study of female ejaculation, with the delightful subtitle "Variations on an Ancient Battle of the Sexes."[28] (It bears mentioning that some women, too, are skeptical of female ejaculation, but more on that later.)

A handful of scientists familiar with the female prostate and ejaculation have translated historical works of erotica or sexology. These include sinologist Rudolf Pfister, Indologist Renate Syed, and physician Karl F. Stifter. How can female ejaculation or squirting be identified in old texts when ambiguous terminology is used to describe bodily fluids? To distinguish this juice from other sexual fluids, it should demonstrate one of the following characteristics. The fluid is emitted from either the vagina or urethra during sex, but isn't urine.[29] Emission occurs right before, after, or during climax, and is accompanied by intense feelings of pleasure. Significantly more fluid is produced, its emergence more abrupt, forceful, or streamlike than vaginal lubrication.

Depending on the source, anywhere from between 10 and 69 percent of women—the latter statistic mentioned earlier—ejaculate during sex.[30] The online erotic portal JOYclub conducted a survey of five thousand female users, nearly 70 percent of whom indicated they had "squirted" before.[31] Despite the high figure, it's likely many women and those who have vulvas don't realize they ejaculate. In heterosexual pairings, a cisgender woman will likely assume that the guy is responsible for soiling the sheets and that any sexual fluid of hers is vaginal wetness. If, however, she feels certain she squirted something, she might further assume—to her profound embarrassment—that it was urine. More on that later too.

Urogenital anatomy often differs significantly between women, as do characteristics of the female prostate and sexual fluids. The prostate's size, shape, and position may vary, while different individuals' ejaculate might be clear or milky, thin or creamy. In some cases, a teaspoon of ejaculate is produced, while in others, a whole cup's worth comes coursing out. The smell and taste of the fluid also varies. Some women and those who have vulvas can identify how their cycle, diet, and stress levels influence the quantity and quality of ejaculate they emit.

Until fairly recently, research examined all such fluids under the term "female ejaculation." In 2011, that changed. Female ejaculation and squirting are now described as distinct phenomena. When climaxing, women and those who have vulvas release two fluids simultaneously or in close succession: a viscous, whitish secretion from the prostate and a watery liquid produced in the bladder. Given the clear differences between these fluids, in 2011 researchers Alberto Rubio-Casillas and Emmanuele A. Jannini recommended using the terms "female ejaculation" for the former, and "squirting" for the latter. The distinction between ejaculation and squirting has been taken up in several scientific publications and slowly seems to be gaining acceptance.[32]

For ejaculation or squirting to occur, where a woman or person with a vulva is stimulated (by themselves or a partner) is irrelevant. Women ejaculate or squirt in response to vaginal stimulation, clitoral excitement, or anal intercourse, and they have "wet dreams" too.[33] According to a 2013 study, 13 of the 320 women surveyed had experienced nocturnal emissions.[34] *Femmes fontaines*—as ejaculating ladyfolk are known in French—and their partners take an unequivocally positive stance toward ejaculation.[35] One happy guy posted on JOY-club's online forum, "What could be better than the feeling of a woman totally giving herself to you, totally trusting you, and then . . . flowing freely in her experience of pleasure."[36]

Nevertheless, more than forty years after female ejaculation was brought to people's attention and popularized by the women's health movement and such books as the international bestseller *The G Spot: And Other Discoveries about Human Sexuality*, most studies these days conclude with an urgent appeal for further research. This book will outline what's been known about vulval juices over past millennia, and the poetry, prose, and pontification it's inspired. Let's roll out the red carpet for sodden sheets, feminine love juices, moist orgasms, and *femmes fontaines*.

1

MULBERRY TREES AND JADE WATER: FEMALE JUICES IN TEXTS FROM EARLY CHINA

THE PLAY OF CLOUDS AND RAIN

Ancient Chinese culture is as heterogeneous as the history of sexuality in China is complex. In this chapter, I outline mostly Taoist takes on sexuality before taking a closer look at the ideas and sexual practices presented in a text dating back more than 2,200 years.

Sex is venerated in ancient China. The erotic encounter and sexual union of man and woman are considered a critical cultural act, body art, medical exercise, and pleasurable release. Heterosexual intercourse reflects universal, cosmic powers. When a couple sleeps together, heaven and earth become one. The term *yün yü*—the play of clouds and rain—is thousands of years old and still in use today to describe amorous romps. The commingling of sexual fluids is celebrated in sexual handbooks that predate Christianity and the erotic tracts of the Middle Ages.[1] Sex far transcends the mere act of reproduction. It's good for both parties' health and brings the sexes into harmony. Carnal desire is at once spiritual desire, with sex amounting to a physical and spiritual therapy that can lead to enlightenment and longevity.

The world is built on two complementary aspects, *yang* and *yin*. Yang is the positive principle, and yin the negative. Yang stands for the heavens, sun, fire, light, summer, and man; yin represents the earth, moon, water, darkness, winter, and woman. Yang and yin correspond to and complement one another, thus creating a living, moving equilibrium that repeats itself in the bodily encounter between men and women. One completes the other; two opposites become one unified whole in this flowing game. Such fluids as saliva, sweat, milk, and sexual excretions circulate between the bodies. Lovers drink one another, inhale the beloved's breath and scent, absorb their energy and power through every opening in the body. Sex is essential to well-being. It stimulates the vital energy chi and may even be the ticket to immortality. A man can die of too little yin.[2]

Men and women engage in sexual intercourse as equals, neither body superior to the other. There is one critical difference between the sexes, however, and it's related to female ejaculation: whereas men have a limited supply of semen at their disposal, women's genital fluids never run dry. Men must therefore control ejaculation, lest the loss of semen weaken and rob them of vitality, whereas women orgasm repeatedly and ejaculate as often as possible. Women are a potent force in this vision of sex and eroticism. They are generous in furnishing their precious fluid and relish their unbounded lust. Ideal sex tends to be slower and more varied, allowing women to climax and ejaculate, and men to replenish their energies and experience pleasure, but only ejaculate if pregnancy is desired. (The five days following the end of menstruation are considered a woman's fertile period.)[3]

What does "ideal" sex look like anyway? How should a man kiss and touch his lover to release her fluids? What's the best way to honor mutual excitement and physical union? And how can couples guard against less-than-ideal sex, the kind in which the man ejaculates, while his partner goes unsatisfied? Chinese bedchamber texts, some written more than two thousand years ago, address such concerns. These ancient sexological manuals, frequently presented as wedding gifts, are a primary educational resource for the upper class. Works like *Wu-Ch'eng-tzu-yin-tao* (*Sex Handbook of Master Wu-Ch'eng*) or *Hé yīn yáng* (*Joining Yin and Yang*) include precise descriptions of the human body, sexual response, and climax as well as commentaries on the health effects of physical arousal and congress. *Tiān xià zhì dào tán* (*Talks on the loftiest ways under heaven*), a text from more than twenty-two hundred years ago, explains that while humans know innately how to breathe and eat, mastery of how to "join man and woman" is gained only after a period of study.[4]

Jīng is one of the most important bodily fluids, though the term applies to any fluid that is whitish or semitransparent, viscous or slimy. Saliva, sweat, and milk are *jīng*, as are male and female sexual fluids. Male ejaculate is also known as *yīng jīng*, which translates roughly as "semen of the concealed parts."[5] A woman's *jīng* essence encompasses all of her sexual fluids: the vaginal lubrication that begins during foreplay and a second fluid that far surpasses the first in quantity, which flows or sprays out of the vulva or urethra right before or during climax—female ejaculation or squirting. This substance has many names: *yīng jīng*, chi, spring water, ambrosia, moonflower water, or peach or melon juice.[6] The man can "drink" (*he*), "consume" (*shih*), or "inhale" (*hsi*) this fluid with his penis.[7]

Every last Chinese sex manual produced from the pre-Christian age into the seventeenth century addresses this essential difference between the sexes. A man's *jīng*, his source of health and vitality, must be administered with utmost care. Sexual encounters can weaken him, making him susceptible to illness and premature aging. It's healthiest for men to practice *coitus reservatus*, ejaculating only when trying for pregnancy. Abstaining from ejaculation enhances health and restores youth. A successful sexual encounter can turn white hair black and make lost teeth grow back. If a man holds in his semen during orgasm, the fluid can climb his spine into the brain, where it triggers "spiritual elucidation" and enlightenment. A man who "slurps up / the finely viscid spirit [male semen] / might then be granted enduring sight / and with heaven and earth attain equal might."[8] Men who do not ejaculate, the medieval writ *Xuang Nü Chin* (*The Classic of the Plain Girl*) promises, will experience a noteworthy boost in strength: "Once the woman feels pleasure, she herself starts to move, and her juices begin to flow freely, whereupon he may penetrate as deeply as possible. When she attains orgasm, they stop. Practice this, without losing [your sperm], and your powers will increase hundredfold."[9]

There's a different quality to women's inexhaustible yin essences. Female fluids nourish and heal men. A woman may masturbate or sleep with other women without censure, the conventional wisdom holding that the more sex she has, the more energy laden her waters—and the greater benefit to her male partner. He goes to great lengths to pleasure her, tap her juices, and prolong coitus, as this allows him to "drink" and "suckle" freely. Other female substances—like breath, sweat, saliva, or milk—also fortify him, though none is as potent as the essence of yin.[10] Sex occasions a reversal of social roles: the penis is considered a "guest" in the woman's vagina. Needy, "*jīng*-poor" men, who envy women their yin, are met with profligate "life donors" when they have sex.[11]

Women are slow to heat up during sex, but like water, they are also slow to cool back down. For that reason, women love "slowness" (*hsü*) and "endurance" (*chiu*), and detest "haste" (*chi*) or "violence" (*pao*).[12] In old Chinese texts, lengthy foreplay is deemed essential, and when vaginal penetration finally does occur, late in the proceedings, it's expected to be varied. Thrusting is by turns shallow and deep, fast and slow. These sex manuals describe upward of thirty positions for penetration, including "path of the tiger," "fighting monkeys," or "reverse dragon." Intercourse is depicted as a game, in which both players employ cunning to gain the other's life-prolonging energy.[13] After all, the woman benefits from her partner's *jīng* too and therefore will try to "steal" it. If she drives him to accidental ejaculation, she wins. There is nothing more perilous than unchecked lust: "When semen announces its approach, hasten from the land! Sleeping with a woman is like mounting a galloping horse with brittle reins, like strolling along the edge of an abyss, whose floor is studded with swords, and into which one fears he might fall. Be sparing with your seed, lest your life draw to a close!"[14] An unsatisfied woman cannot be contained, and when left wanting by marital sex, she may avail herself of more competent lovers, thus cuckolding her husband—as even pre-Christian texts demonstrate. Conversely, if a woman is sexually fulfilled, coupledom can bring great pleasures. It is well worth it for men to be good lovers, practicing and perfecting endurance, physical control, attentiveness, empathy, and technique.

For couples to attain harmony in intercourse, men must learn to read women's longing. Early Chinese sex manuals paint a complete picture of female sexual response and provide precise instructions on how men should react to each cue. What can one glean from his lover's groans and sighs, or the flush of her face, curl of her toes, lift of her pelvis, or scent and stream of her fluids? Is now the right moment to caress her arms? Does she prefer deep insertion or shallow, fast movement or slow? Should they change position or maybe even pause? The female body is closely observed, with the color, size, and depth of women's genitals—and the fluids they produce—described in poetic detail. These texts map out parts of the vagina known today by medical names alone. What about an erotic term for the area to either side of the cervix, though? Or a word to indicate an awareness of the vaginal walls, beyond the G-spot (see chapter 5 for a discussion of this term as well)? These are the "jade gate," *yumen*, and "red room," *zhushi*, respectively.

The Mawangdui texts on the arts of the bedchamber, two bamboo manuscripts written more than twenty-two hundred years ago, are "among the earliest surviving accounts in the world of such detailed sexual knowledge," and provide insight into ancient Chinese views on bodies and sex.[15] In translating and studying these texts, Swiss sinologist Rudolf Pfister uncovered references to two phenomena that would prove contentious more than two millennia later: female ejaculation and the G-spot.

MILK FRUIT AND RISING CHI: THE MAWANGDUI BAMBOO MANUSCRIPTS

In 1973, three pre-Christian burial sites are discovered in Hunan Province in China. The luxurious graves of the Li family contain silk clothing, musical instruments, cosmetics, baskets of food, lacquerware—and a library. Of the twenty surviving manuscripts, which include various medical works, two focus on sex. The characters preserved on these bamboo slips represent the oldest-remaining artifact from China to "focus on the female body in such detail."[16] The anonymous authors, some of whom may well have been women, craft concise yet tender descriptions of positions, movements, sounds, and fluids.[17] The texts address female ejaculation and squirting in one of the earliest-known references to the matter.

The first manuscript, *Talks on the Loftiest Ways under Heaven*, consists of fifty-six bamboo slips, whereas the second, *Joining Yin and Yang*, is slightly shorter, inscribed on thirty-two slips. Both depict sex as a celebration of the bodies, and no detail is spared in describing how men should read and respond to women's arousal: "First, the *qi* rises, the face heats up; slowly kiss. Second, the nipples stiffen, the nose moistens; slowly embrace. Third, the tongue becomes thin and slippery; slowly draw closer. Fourth, waters flow between the thighs; slowly caress. Fifth, the throat dries and she tries to swallow saliva; slowly rock her."[18] Men are instructed to mount their lover only once the "Five Desires" have manifested.[19]

The vulva and vagina are portrayed with similar, though in some cases indecipherable, attention to detail. *Talks on the Loftiest Ways under Heaven* lists twelve names denoting parts of the female sexual anatomy, most of the terms now opaque: hair clasp glossiness, bulging mark, split gourd, mouse or wood louse, grain fruit or milk fruit, wheat teeth, enclosing clothes, recessus, carrier or levator stricture, red rag or shred, red orifice, and pebble stone.[20] The parts described include the clitoral crown, labia majora and minora, fourchette (the spot at the bottom of the labia minora, where these inner lips meet), cervix, and hymenal edge. Both texts depict the erotic charge of lovers drawing close, the role of the breath, kissing, touching, fingering, and "stimulating," the expressions and sounds of female arousal (gasping, whining, and groaning), her movements during sex, and vaginal penetration in all of its variations. The section on female climax reads: "The manifestations of the stage of great completion are: moist air comes from the nose, the lips turn white, the hands and feet move erratically, the buttocks rise off the mat. When intercourse is finished, she hardly breathes (?), as if she were dead. At this time, the *qi* from the inner extreme billows forth, energy gathers within, and thus she attains spiritual clarity."[21] Female ejaculate—qi (chi) spreading from the middle pole—is thoroughly described, from its consistency and color to its smell and taste. It is pure and fresh, porridge-like, creamy, ropy or viscous, smells of fish or grain, and tastes brackish.[22] Amazingly, the Mawangdui manuscripts and other early Chinese texts reference an erogenous zone no more than a couple inches in size that will become one of the most controversial parts of female anatomy in the twentieth century: the legendary Gräfenberg—or simply G—spot. The Chinese authors chose the pom-pom-like fruits of the paper mulberry (*Broussonetia papyrifera*) to portray this area of the vagina. Pfister explains,

> The "milk fruit"—a globular, red-orange cluster of individual fruits encased in flesh—was a fitting floral analogy. Photographs of the anterior vaginal wall reveal delicate fleshy protuberances that swell and darken in color during arousal. These are compared to raspberries (*Rubus idaeus*), another commonplace form that was more button-like. The ripe fruit of the paper mulberry tree exhibits the same structure and juiciness as the raspberry, and both are vibrant red in color: these very characteristics were linguistically applied to the Gräfenberg spot.[23]

The image of a swelling, spraying milk fruit appears in other Chinese works not nearly as old as the Mawangdui bamboo manuscripts. In the medieval text *Dong Xuan Zi* (*Master of Grotto Darkness*), a man stimulates his partner's milk fruit with his *yáng* tip until she trembles. *Su Nü Miao Lun* (*The Wondrous*

Treatise of the Plain Woman), a late imperial work, illustrates how "her red ball" is teased to the point of release. The female partner is instructed to extend her left leg loosely and bend her right, while the man crouches behind and works the "jade"—her vagina—as he likes. He is meant to tap his partner's red "pearl" and penetrate her deeply seven times, followed by eight shallow thrusts. When the woman's red ball grows, it is said to move swiftly and spray. The man then draws her ejaculate into his penis—a method that resembles a golden cicada clinging to a tree and imbibing dew. The cicada holds the dew in its mouth and does not spit it out.[24]

Along with depictions of the G-spot, descriptions of female sexual fluids are common to ancient Chinese bedchamber texts and erotic manuals written in subsequent centuries. These works document female secretions as a matter of course. A handbook from the Sui dynasty (581–618 CE) states that one must approach sex as a game of ping-pong, a volley between action and reaction in which moistness gives way to wetness. If her vagina grows moist, he may slowly penetrate his partner more deeply. If her juices drip down her buttocks, he knows that she's enjoying his skills in the art of lovemaking. The *Hsiu-chen-yen-i*, a popular work from the late Ming dynasty (1368–1644), outlines the health benefits of female fluids, of which there are three: saliva, a fluid produced by the breasts, and vaginal secretions. During sex, the man should drink of this "excellent medicine of the three mounds." The fluids sharpen his wits, boost his vital force, and renew his blood. The medicine of the lower mound—the vulva—is understood to come from inside the vagina. When a woman is aroused and her cheeks flush, the gates of the vagina open and her waters flow, collecting inside her vagina, where the man receives the medicine with his penis.[25]

Ejaculating or squirting women can also be found in early Chinese tales beyond the bedchamber texts. Not only do women's fluids provide nourishment and healing, they're a clear sign of desire and relaxation. How are milk fruit, jade water, and melon juice depicted in erotic literature?

LIKE CARP PLUNGING THROUGH MUD

The great novels of the Ming era demonstrate the significance of eroticism and sex in early China. *Chin P'ing Mei* (*The Plum in the Golden Vase*) and Li Yu's *Rou Pu Tuan* (*The Carnal Prayer Mat*), for instance, spare no details in portraying their protagonists' erotic exploits. Female sexual secretions feature prominently in both novels, and here too, a distinction is made between two vaginal fluids: lubrication, which accompanies the start of the amorous encounter, and a second fluid that flows or sprays liberally at the point of climax.

Chin P'ing Mei, composed in the latter half of the 1500s, is considered one of the great classical novels in Chinese literature. Although the story is set in the twelfth century, the novel is understood to depict life during the late Ming dynasty, thus serving as a key sociocultural artifact of the era. The hundred-chapter book follows the life and amatory adventures of a pharmacist and silk merchant in Shandong Province. Hsimen Ch'ing is a wealthy man. He has six wives and maintains countless extramarital relationships. Fluids gush and streams unite in well over a hundred graphic sex scenes. *Yün yü*, the play of clouds and rain, could be used to describe this passage: "The ruttish cry soon issued from her mouth like the hoarse shriek of the cockatoo, while he diligently mimicked the bright butterfly that lustfully dives into the sweet depths of a fragrant flower. Two reconciled spouses found refreshment that night in the quenching wetness that streams steadily from the gathered clouds." Female fluids are celebrated. In this case, the penis does not consume them, cicada-like, but rather swims in them like a fish: "[Hsi-Men Ch'ing] continued to thrust until her vaginal secretions flowed uninterruptedly, making a sound like a school of loaches [carp] plunging through the mud."[26] *Chin P'ing Mei* also illustrates that women's elevated position in the bedchamber is at odds with their standing in society, where they are subordinate to men, and live with the threat of sexual exploitation and violence, particularly when they aren't upper class. The novel tells that side of the story as well.

In *Langshi*, we witness the sexual escapades of playboy Mei Shusheng unfold over the course of forty chapters. One day, this man-about-town meets the beautiful Li Wenfei in a cemetery. It's love at first sight. They get to chatting and end up spending the night together. We are told how, nearing climax, Wenfei screams with passion, and her "dew," which dampens the bed beneath her, is no longer white; it is clear as albumen and then takes on a reddish shimmer. The two go on and on, and their secretions flow in streams into a small lake under their bodies.[27]

An erotic novel from the seventeenth century, *Ge-Lian Hua-Ying* (*Flower Shadows on a Window Blind*), portrays a young lesbian couple making love. During the young women's nightly play, the vulva of one grows very wet; her lover takes the fluid to be urine, only to learn from her more experienced girlfriend that "this is women's sexual secretion, and tomorrow when I play with you, you will get as wet as I am."[28] The confusion between sexual emissions and urine is unusual for this literature, because early Chinese sex manuals and erotic novels are clear about which of these liquids feature exclusively during sex. In this instance, it's a sexually inexperienced teenager asking if her lover peed the bed. In the nineteenth and twentieth centuries, a sense of uncertainty regarding ejaculate and urine will take root, coupled with women's fear of accidentally tinkling during sex. European and North American medicine and sexology, in particular, will habitually misinterpret vulval juices as incontinence, to the point where many modern-day *femmes fontaines* unfortunately still approach their emissions with trepidation.

2

WATERS OF LUST AND LOVE: FEMALE JUICES IN ANCIENT INDIAN TEXTS

Female fluids are known in India as well, where myriad historical texts portray their release during sex as a fundamental aspect of female pleasure. One Sanskrit term for ejaculate, whether male or female, is *śukra*.[1] In one of the oldest-surviving accounts of female ejaculation in Sanskrit literature, seventh-century CE poet Amaru (or Amaruka) describes lovers in close embrace, a shudder passing through the woman's body as sweet *śukra* flows onto her raiments.[2]

Hinduism outlines three aims in life: *artha*, or material prosperity and success; *kama*, or worldly pleasure, desire, and sexuality; and *dharma*, or the law, justice, morality, and ethical obligations.[3] Sexual pleasure and eroticism are central to human experience.

The pursuit of *kama* is both theoretical (one reads about sex) and practical (one has sex), which gives rise to *kamashastra*, an Indian literary tradition dedicated to the knowledge of erotic love. The most famous book in the canon is the *Kamasutra*, by Vatsyayana Mallanaga. This manual, however legendary, does not explicitly describe female ejaculation or semen, but references to both are ubiquitous in later works of ancient Indian sexology. Secretions are part of female pleasure, driving women to the brink of insensibility. And these fluids are virile; in the case of the Tantric goddess Kubjika, they create the universe itself.

Ancient and medieval kamashastra texts are a trove of insight into carnal love. As in China, sex in India is not defined by mindless passion. Rather, it's a codified physical and spiritual practice, its tenets studied, taught, and put into action. Erotic textbooks like the *Kamasutra* are rich compendiums of these techniques that serve as inspiration for Tantrists and other later writers. From Vatsyayana's original *Kamasutra* (which has seen many subsequent commentaries and addenda) to other manuals, a literature emerges in India that celebrates the erotic, tenderness, and intercourse. Authors within this tradition are considered early "flag bearers for sexual pleasure." Kamashastra writings demonstrate that "pleasure needs to be cultivated, that in the realm of sex, nature requires culture."[4] Detailed descriptions of sexual acts are intended to acquaint readers with the body of the other, thus increasing intimacy between lovers.

Vatsyayana composes the *Kamasutra* between the third and sixth centuries CE.[5] One thousand years or so later, Yashodhara Indrapada publishedes the best-known commentary on the erotic textbook, the *Jayamangala*. The thirteenth-century poet opens his remarks by stating, "Those who talk about pleasure say that pleasure is the most important of the three aims of human life, because it is both the cause and the result of the other two, religion and power. And realizing that nothing happens without a method to make it happen, and wishing to explain that method, the scholar Vatsyayana Mallanaga created this text."[6] Unlike medical writings of the time, kamashastra texts discuss sex in terms of enjoyment, not procreation. The guiding principle of these manuals is pleasure, down to descriptions of anatomy and genital functions. Sex is a thing to be enjoyed by both men and women. Men, however, are responsible for women's gratification, and

Figure 2.1
Ejaculating or squirting woman waterspout on an Indian temple.

expected to "act as a paragon of gentleness and consideration. His every thought must be for his partner's pleasure and satisfaction. Although the Indian woman is in society totally subject to masculine control, in love she is entitled to claim from the man every conceivable indulgence. Pleasure and fulfilment are her right, no less if she is courtesan or prostitute, than if she is a wife. . . . She it is who must be satisfied."[7]

If responsible for his partner's sexual enjoyment, a man must know his lover as well as possible, know her traits, peculiarities, proclivities, needs, and habits. To that end, kamashastra scripts offer a tidy taxonomy. Men and women are classified by looks, age, social standing, temperament, home province, origins, and so on. There are encyclopedic entries on sexual positions, kissing techniques (where to kiss and how, pecking or prodding), manners of embrace (touching, penetrating, rubbing, or pressing), scratching, biting, hairpulling, slapping (how, where, and when), and moaning. The works close with recipes for aphrodisiacs, cosmetics, and magic potions.

Most important among the kamashastra organizing principles, though, are the genitals, namely the size of the penis as well as the depth and width of the vagina. Certain pairings provide little pleasure (small penis, deep vagina), whereas others are extra enjoyable (medium-sized penis, narrow vagina). If a man and woman are well suited, and he is practiced in erotic arts, the man will get his partner to orgasm and ejaculate or gush.[8]

LUSH GUSH: FEMALE FLUIDS IN KAMASHASTRA TEXTBOOKS

The texts identify two genital fluids women produce during sex: *syandana*, or vaginal lubrication that sets in with arousal, and *viṣrsṭi*, the surging effusion that coincides with climax.[9] The fluid women ejaculate is known as *retas* or *śukra*, meaning semen, the same term used for male ejaculate.[10] Yashodhara mentions these fluids in the *Jayamangala*, clearly differentiating between the two:

As it is said:

> A woman's sensual pleasure is twofold:
> the scratching of an itch and the pleasure of melting.

The melting, too, is twofold:
the flowing and the ejaculation of the seed.
She gets wet just from the flowing,
and her sensual pleasure of ejaculation comes from being churned.
But when a woman is carried away by her sexual energy,
she ejaculates at the end, it is said, just like a man.[11]

Indian scholar and poet Kokkoka writes the classic medieval sex manual *Ratirahasya* between the eleventh and twelfth centuries. He, too, differentiates between two fluids:

With the admission of a firm phallus into the woman's yoni, she experiences the slow but steady disappearance of the uncomfortable itching sensation, but when the oozing begins, she experiences pleasure, and her pleasure increases in proportion to the free flow of her fluid.

In the beginning of the union between man and woman, there is a feeling of pain, resulting in meagre pleasure. Both the man and the woman experience pleasure only towards the end.

In the end, the woman utters indistinct and hardly audible sounds, writhes her body, feigns crying, gets extremely perturbed, and closing her eyes, reaches a state of helplessness and lassitude.[12]

Kamashastra texts distinguish between and classify women according to the appearance, internal qualities, and smell of their genitals—or *yoni*—as well as sexual fluids. In *Ratirahasya*, Kokkoka writes of four types of women, "Padmini, Chitrini, Shankhini and then Hastini." He enumerates first the superior characteristics of the Padmini, whose "yoni is cup-shaped, like a lotus in bloom," and whose "mucous discharge too has the unusual fragrance of a blossoming lotus." Meanwhile, the Chitrini, whose "mucous discharge has the fragrance of honey," is also known by her "large breasts and yoni," which is "well-rounded and high, soft and well-lubricated inside." The "quick-tempered" Shankhini, "fond of red flowers and garments," produces an acidic-smelling fluid. Finally, one can recognize the Hastini by her lumbering gait, "crooked toes," and "excessive hairiness." Her "yoni and mucous discharge smell like the rut of an elephant." Intimate smells and secretions will also differ between regions of India. Women from Dravida, for instance, "have a profuse flow of fluid, and are much excited by the introduction of the finger into their yonis and by the exterior acts of kissing and so on. They attain their climax quickly during the initial congress."[13]

One of the most important erotic textbooks of the late Middle Ages is commissioned by an Islamic ruler in northern India who enlists poet Kalayana Malla to write a manual for married men demonstrating that monogamy could be sexually fulfilling. In *Ananga Ranga*, translated into English in 1885 by Sir Richard Francis Burton, Malla explains how to pleasure women, as marital harmony depends on both parties' sexual satisfaction. The author provides detailed instructions on foreplay and introduces a series of sex positions. He too devises typologies of the sexes. As in *Ratirahasya*, the four types of women—delicate lotus-like Padmini, coquette of the arts Chitrini, bilious Shankhini, and pachydermal Hastini—differ in temperament, behavior, skin, eyes, voice, and other physical traits.

The typology in *Ananga Ranga* includes the appearance of the yoni and nature of its fluids. *Kama-salila*, Sanskrit for the female "love juice," is the same word used for male ejaculate (though not for male semen).[14] The yoni of the beautiful Chitrini is "soft, raised, and round," and her seed is "hot, and has the perfume of honey, producing from its abundance a sound during the venereal rite." The tall, small-breasted Shankhini, on the other hand, produces "distinctly salt[y]" ejaculate. The Hastini's "*kama-salila* has the savour of the juice which flows in spring from the elephant's temples."[15]

One of the many sex positions presented in *Ananga Ranga* features female juices and addresses women's right to sexual satisfaction:

> *Purushayitabandha* is the reverse of what men usually practise. In this case the man lies upon his back, draws his wife upon him and enjoys her. It is especially useful when he, being exhausted, is no longer capable of muscular exertion, and when she is ungratified, being still full of the water of love. The wife must, therefore, place her husband supine upon the bed or carpet, mount upon his person, and satisfy her desires.

Unlike ancient Chinese tradition, in historical Indian texts ejaculation is part of orgasm for both lovers—men included. Women welcome unhurried sex and deft erotic play. An experienced male lover seeks to heighten his partner's pleasure, getting her to orgasm and ejaculate before coming himself.[16]

German Indologist Richard Schmidt (1866–1939) translates the *Kamasutra* into German in 1900. In his oft-reprinted publication *Beiträge zur indischen Erotik* (Essays on Indian eroticism) from 1902, he introduces curious readers to the "love life of the Sanskrit-*Volk*," as the subtitle reads. Schmidt, a professor of Indology at the University of Münster and longtime lecturer at the University of Halle, renders a host of Sanskrit source materials in German, most notably the erotological kamashastra texts. He translates, collates, and cites recipes and magic spells. The "esoteric doctrines" (*upanishad*) Schmidt gathers include acne treatments and breath fresheners, tips for delaying male ejaculation, and ways to overcome challenges related to female orgasm. For instance, a woman need only apply an unguent of tamarind peel and honey to the inside of her vagina to prompt her seed to flow sooner than that of her partner. Beautiful women are quick to express their seed in congress with a man who has rubbed an ointment of *mahesa* (possibly giant milkweed) seed, borax, honey, and camphor on his penis.[17]

A SWELLING LITTLE TUBE AND FLOOD OF WATERS

Kamashastra manuals employ a purple patois to name and describe the vulva and vagina. How else could writers portray them in their full plurality? Kokkoka's *Ratirahasya*—a book so popular it was translated into Arabic, Persian, Turkish, English, and German—explains it like this: "The yoni is of four types. It can be soft like a lotus inside; or it can be taut, like fingers held tightly together; or it can be slightly wrinkled; or it can be like the tongue of a cow."[18]

The texts also mention a spot of particular relevance to the discussion of female ejaculation or squirting. There in the middle of the vagina, kamashastra guides read, is a little tube that swells and unleashes a "flood of lust water" when stimulated with the phallus.[19] The authors encourage digital play as well; the *Ratirahasya*, for instance, mentions "a nerve center, resembling a phallus, in the center of the vagina, which is known as *Madanagamanadola*. When this is manipulated with two fingers, it generates the flow of the woman's fluid."[20]

This phallic little tube provokes a great deal of head-scratching among later Indologists like Schmidt, who gets a physician to review these depictions of female sexual anatomy. The doctor's "kind efforts" show "that most descriptions provided by the Indian authors are incorrect, and that the organs they refer to inside the vulva [*sic*; Schmidt means vagina] have proven impossible to identify."[21] Nearly a century later, in 1999, Munich based Indologist Renate Syed reaches a much different conclusion in her own extensive, fascinating study: the vaginal area that appears in these old Indian books is none other than the G-spot, in today's nomenclature. Like the G-spot, the little tube described in the texts swells, hardens, and stands erect, thus resembling "the Indian conception of the penis." Women who enjoy this form of stimulation will emit a fluid that differs from vaginal lubrication and that the author identifies as female ejaculation. The old Indian writers, Syed avers, were familiar with "the phenomena of the 'Gräfenberg zone' and female ejaculation."[22] The erotic writers and poets of ancient India and China—those introduced in the last chapter who called the G-spot "milk fruit" and detailed its swelling and attendant ejaculation—were therefore far better informed about the female body than venerated old Professor Schmidt centuries later.

BRIEF ASIDE: FEMALE EJACULATION IN TANTRA, PART I

Women and their bodies are revered in Tantric Buddhism, a major religious tradition across Asia since the Pala Empire (eighth–twelfth century). Practitioners render homage to female Buddhas and founding mothers. They even have their own creation myth. The Tantric goddess Kubjika—also known as Śukrādevī, literally "Goddess Sperm" —is a cave-dwelling recluse.[23] Weary of her ascetic existence, she licks her own vulva. This acrobatic feat is due in part to Kubjika's name, which means the bent or hunchbacked one.[24] The goddess climaxes and ejaculates or squirts. Her seed spouts from her vulva, the center of her yoni, her mandala.[25] The creation of the universe can thus be traced back to masturbation and female ejaculation or squirting.[26] The Sanskrit term *yoni* means source or blessed place, the seat of emotion. It signifies both internal and external female genitalia, and represents divine female energy. The yoni and its sexual fluids are sacred, honored in such Tantric rituals as *maithuna* or *yoni puja*.[27] Men and women historically share in the discipline of Tantric philosophy, which is notable in its appreciation of the female body. Women are active practitioners and teachers, equal to men. Tantric texts emphasize that women should be treated with respect and honored in rituals. The ideal male-female relationship is based on partnership, cooperation, and equality:

> One should honor women.
> Women are heaven, women are truth,
> Women are the supreme fire of transformation.
> Women are Buddha, women are religious community,
> Women are the perfection of wisdom.[28]

Enlightenment is the goal of religious practice for men and women alike, and attained through meditation, yoga, and *mudrā*, or symbolic hand gestures. That, and sex. Ritualistic intercourse, in which both lovers orgasm and ejaculate, stimulates sexual and spiritual energies. The longer the act, the more effective. Sexual congress in Tantra is passionate and spiritual. Breath, fluids, and energies move between the bodies, these elements renewed by the exchange. The man controls his erection, orgasm, and ejaculation, and stimulates the flow of fluids in his female partner. As religious scholar Miranda Shaw explains, both engage "the energies and fluids circulating through one another's bodies to become enlightened beings."[29]

Enlightenment is the hope in making love and mixing fluids—not procreation. Ejaculate, whether male or female, is known by the term *śukra*. Female sexual fluid is also called *madhu* (meaning something sweet, like honey, wine, or nectar), "flower water," "musk of desire," or "musk of intoxication."[30] Old Tantric texts sometimes refer to female ejaculate as red. The very mention of red has led to myriad errors in translation and the erasure of female ejaculate, rendered instead as menstrual blood. During Tantric intercourse, male and female sexual fluids mix:

In the sacred citadel of the vulva of
A superlative, skillful partner,
Do the practice of mixing white seed
With her ocean of red seed.
Then absorb, raise, and spread the nectar, for
A stream of ecstasy such as you've never known.
Then for pleasure surpassing pleasure,
Realize that as inseparable from emptiness.[31]

Sexual fluids have a particular power: they nourish the spirit, meaning that after mixing, the fluids are absorbed by the penis (*vajroli-mudra*) or vagina, or ingested orally: "This is the best diet, eaten by all Buddhas."[32] There are many Tantric rituals in which female juices are drunk straight or combined with male semen or wine.[33] Although sexual fluids are mixed and consumed by both parties, Tantric Buddhism is also guilty of placing more emphasis on men's absorbing the female essence. In her groundbreaking study of women in Tantric Buddhism, Shaw writes that the *Hevajra-tantra* urges men "to vow to continue to absorb the female essences in all future lives, until enlightenment is reached."[34]

Tamil Tantra texts describe a female sexual fluid (*tiravam*) analogous to sperm (*vintu*) with spiritual, medical, and therapeutic qualities. Male semen and female emissions are collected, blended, and presented as offerings to the gods. Drinking this concoction is believed to improve male Tantrists' virility and health. The fluids are even fashioned into pills and used to treat mental illness.[35] The medieval Tamil poem *Kāmapānacāstiram* (Treatise on the arrow of lust) depicts the act of cunnilingus. The text celebrates the yoni swelling and growing moist as it's licked, kissed, sucked, and nibbled until it bursts with fluid. The reader is invited to "inhale deeply, breathing in the mellow odours of the juices of her yoni" because not only do they smell good, when mixed with male semen, but these fluids are an invigorating drink too.[36]

In the 1960s, Tantra becomes a touchstone of some Western liberation movements, inspiring hippies, feminists, and New Age adherents with the value it places on women. The practice celebrates the female body, maintains the equality of women and men, and proposes a paradigm for the nonhierarchical encounter of bodies during sex. Western Tantra pioneers like Margot Anand or Caroline Muir and Charles Muir popularize the practice, sharing centuries-old knowledge about female sexual emissions in workshops, books, and videos.

The Indian sex manuals and Tantric texts introduced here address aspects of a female genital fluid related to sex, pleasure, and orgasm, but not reproduction. The notion of a woman "fathering" a child also exists in old India. The belief that babies are made by a mixture of female and male semen extends well into the pre-Christian era, as evidenced in the sacred Hindu text *Rigveda*, written in the latter half of the second millennium BCE, or the *Taittirya Samhita* from circa 800 BCE.[37] There are various takes on how male and female semen form the fetus. Suśruta, widely recognized as India's first surgeon, maintains that a preponderance of male sperm will produce a male child, whereas the baby will be a girl if more female sperm (*ārtava*) is present at conception.[38] A child of the third gender, or *hijra*, may result from equal parts male and female sperm.[39] Another ancient text offers this explanation for how a baby's sex is determined: on even days, a woman's seminal substance is so negligible that a boy will emerge from the marriage bed; on odd days, that seed is stronger, resulting in a daughter.[40] Such numbers games are a thing of the past, but even in today's India, the idea of a generative female fluid remains—*rāja*.[41]

3

FIRST COMES PLEASURE, THEN PROCREATION: THE FEMALE SEED FROM ANTIQUITY TO THE MODERN AGE

Conceptions of the female body in ancient Greece and Rome contrast significantly with those presented in historical Chinese and Indian texts. Women's sexual desire and satisfaction are largely overlooked, and their sexual fluids are depicted much differently than in old Asian texts. Ancient Greek and Roman philosophers and physicians don't portray female juices as a coveted fluid a man tenderly coaxes from his partner in amorous play, nor as a nourishing, life-prolonging substance a woman generously bestows on her lover. The female seed is now discussed primarily in the context of reproduction and embryology. Gone are the instructions on how to stimulate your ladylove to the point of orgasm and ejaculation or squirting; the focus has now turned to the part her seminal fluid plays in forming the unborn child.

Ancient Greek and Roman notions of female semen encompass a range of sexual fluids; by no means do these texts all describe ejaculation or squirting. To wit, the legendary Hippocrates of Kos (fifth–fourth century BCE) writes that women sometimes ejaculate into the uterus—an anatomical impossibility.[1] Countless medical and philosophical tracts do, however, mention a fluid bearing a likeness to female ejaculate or squirting. Although ejaculation as we define it today isn't exactly congruous with the classical understanding of female semen, the phenomenon *is* one aspect of it.

The feminine is not elevated in ancient Greece and Rome. Men and women differ by degrees, not in essence. In *Making Sex*, his study of the construction of gender, historian Thomas Laqueur writes,

> The one-sex body would seem to have no boundaries. . . . There are hirsute, viral women . . . who are too hot to procreate and are as bold as men; and there are weak, effeminate men, too cold to procreate and perhaps even womanly in wanting to be penetrated.[2]

Women are generally considered a flawed version of men, passive creatures inferior in anatomy, physiology, and psychology.

For more than two millennia, Laqueur tells us, from antiquity to the eighteenth century, the European conception of

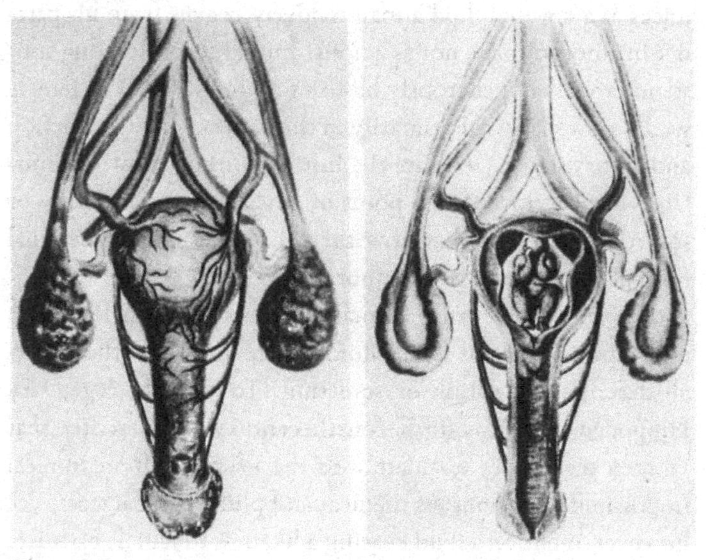

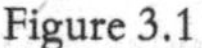

Figure 3.1
The left-hand image, from Georg Bartisch's *Kunstbuche* (1575), depicts the penile female reproductive organs. The image on the right provides a cross section of the uterus.

male and female bodies is that of one body, men, representing the standard and really the ideal. Women have the same reproductive organs as men, only theirs are on the inside: the vagina is an inverted penis; the ovaries are testes that remained in the body.

Medical terminology reflects this belief. Ancient Alexandrian physician Herophilus, a prominent early anatomist, is the first to describe the ovaries, drawing a clear analogy to the male body by referring to them as "testes." This is where female semen is produced before it travels along "seminal ducts" to the bladder and leaves the body through the urethra. The ovaries won't have their own name for thousands of years.[3]

FEMALE SEMEN AS A GENERATIVE SUBSTANCE

As fluids course through the body, they are altered, one turning into the other. Male and female bodies produce and process these substances differently: "What one sees, or could ever see, does not really matter except insofar as the thicker, whiter, frothier quality of the male semen is a hint that it is more powerful, more likely to act as an efficient cause, than the thinner, less pristinely white, and more watery female ejaculate or the still red, even less concocted, menstrua," writes Laqueur.[4] Semen fits into the concept of a comprehensive physical symmetry (or rather, asymmetry) between the sexes. Men emit the fluid at the height of arousal, as do women—some from their urethral opening, and others from their vagina. Some women express far more semen than men, and most only during climax. Philosophers and physicians, however, simply can't agree on what female semen looks like. Some describe it as milky, and others as clear. Sometimes it's thin, and other times it's viscous. We encountered similar variance in depictions of vulval juices in the old Chinese and Indian texts. The described differences of juices fit well with the current debates that propose a distinction of fluids into "ejaculation" and "squirting."

There is a long tradition of belief in female generative power, even in patriarchal societies, given that children resemble both their father *and* mother. Centuries before egg and sperm cells are discovered, Greek natural philosophers and doctors espouse what is now called the two-seed theory, whereby two reproductive substances exist, namely a male seed and female seed. Proponents of the two-seed model offer a range of theories on the seeds' respective procreative properties and influence on offspring.[5] In the fifth century BCE, the pre-Socratic Greek philosopher Anaxagoras states that female seed determines the child's sex and appearance. Alcmaeon of Croton (sixth–fifth century BCE), a pupil of Pythagoras, explains that the fetus inherits its sex from either the mother or father, depending on which parent contributes quantitively more seed. Physician Diocles of Karystos (fourth century BCE) contends that conception will fail if too little female seed is present: if a woman wishes to become pregnant, she must ejaculate.[6]

According to Hippocrates of Kos, who writes extensively about both seeds, a woman's seed will sometimes stream into her uterus and stay there. It may also flow from the uterus into the vagina. A child can only be conceived in the first instance—that is, when female and male seeds commingle in the uterus. Unlike Aristotle in a later age, Hippocrates doesn't see a qualitative difference between male and female seed, maintaining that the substance isn't tied to the biological sex of its originator: "In both a woman and a man there exist both female and male seed."[7] Hippocrates uses this theory to explain why, for instance, a couple can produce both sons and daughters. A child will more closely resemble the parent who pitches in more seed. If a tot looks like its mother, it's because her seed is more "manly" and asserted itself over the father's weaker, more "womanly" seed.

Hippocrates acknowledges that women have wet dreams too. (The phenomenon of unintended nocturnal emissions will take center stage in debates on such neurotic disorders as hysteria around 1900.) Many ancient Greek and Roman physicians describe the gush of semen as a pleasurable experience for men and women alike. Hippocrates also observes that, following ejaculation, women don't experience "as much pleasure" as men.[8]

Several physicians note that women "spill" their seed from the urethral opening. It must not factor into reproduction, then, because how would it get back to the uterus, where the baby gestates? Herophilos of Chalcedon (ca. 325–255 BCE) and Hippon (fifth century BCE) cite this observation to justify their rejection of the two-seed model. Soranos of Ephesus (ca. 98–138 CE), a foremost gynecologist of antiquity and author of *Gynaecia*, thinks that female seed travels from bladder to urethra and then exits the body through the urethral opening during intercourse, meaning he too is convinced that female seed doesn't contribute to procreation.[9]

The most influential critic of the two-seed theory, however, is Aristotle (384–322 BCE). The Greek naturalist and philosopher minimizes women's role in reproduction. Women are doubly deficient creatures in Aristotle's estimation. For one, they're actually men, albeit incomplete men who didn't develop fully in the womb. Secondly, they are unable to produce fertile seed. The two seeds are not analogous. Instead, Aristotle sees a correspondence between menstrual blood, which he calls "unfinished seed," and the male fluid. Male seed alone has the power of reproduction. So what *do* women contribute to baby making? The womb, where the embryo comes into being, and menstrual blood, which provides the embryo its substance: "As principles of generation one might reasonably propose the male and the female, the male as possessing the generative and motor principle, the female as possessing the material principle."[10]

EJACULATION, WET DREAMS, AND BLUE BALLS

The Aristotelian single-seed theory massively diminishes the relevance of female seed. What's interesting, though, is that the philosopher also describes a second fluid many women expel during sex. No creature on earth produces two spermatic secretions, Aristotle reasons, so this fluid must be something else altogether:

> There are some who think that the female contributes semen during coition because women sometimes derive pleasure from it comparable to that of the male and also produce a fluid secretion. This fluid, however, is not seminal; it is peculiar to the part from which it comes in each several individual; there is a discharge from the uterus, which though it happens in some women does not in others. . . . Where it occurs, this discharge is sometimes on quite a different scale from the semen discharged by the male, and greatly exceeds it in bulk.[11]

In Rome more than four hundred years later, Greek physician Claudius Galenus revives the two-seed theory. Galen, the most influential medical figure in antiquity, states that male and female reproductive organs are structurally identical. Women are men "in whom a lack of vital heat—of perfection—had resulted in the retention, inside, of structures

Figure 3.2
Diagram by Andreas Vesalius (1514–1564), from a 1586 publication by Juan Valverde de Amusco (1525–1588). To the left is what looks like a penis, and to the right is the female body from which it was taken.

that in the male are visible without."[12] Women are endowed with the same parts as men, only in different spots: "Consider first whichever parts you please, turn outward the woman's, turn inward, so to speak, and fold double the man's, and you will find them the same in both in every respect."[13] This genital symmetry extends to sexual fluids as well: like men, women produce and secrete fertile seed, though theirs is less perfect or mature by comparison. Galen believes male seed is insufficient for reproduction on its own and needs supplementing: conception occurs only when both seeds emerge at once and commingle. Galen's anatomical studies reveal the ovaries, or inverted "female testicles," as he understands it. Animal dissections reveal a mucous white substance in the female testicles' excretory ducts—a fluid he calls "female seed."

Galen depicts female ejaculation in *Opera omnia*:

> However because the woman is colder than the man, the fluid in her "prostate" (*parastatis glandulosis*) is underdeveloped and thin and thus does not contribute to the creation of life. It stands to reason, then, that it would spill outward. . . . The following markers demonstrate that this fluid not only encourages coitus, but also sparks lust and wets the passage as it leaves: it evidently flows from the woman, spilling perceptibly onto the man's privates when both achieve the greatest pleasure in intercourse.[14]

Galen also describes a phenomenon that afflicts men and women alike, though in modern parlance the term is decidedly gendered: blue balls. According to the ancient Greek physician, when a woman's seed gets trapped in the body, it clots or stagnates. It blocks the healthy circulation of fluids and triggers an impressive array of symptoms. And what does the doctor order in such cases? That the patient touch and warm her genitals. Prescribed masturbation leads to "twitching accompanied at the same time by pain and pleasure after which she emit[s] a turbid and abundant sperm." Ejaculating the stale seed frees the woman "of all the evil she felt" and catapults us into the Middle Ages—an era in which scholars explore female semen in detail, and doctors, midwives, and wise women wield an arsenal of correctives and therapies for seminal blockage.[15]

THE MIDDLE AND EARLY MODERN AGES

Medieval theoreticians adopt the idea of a female seed and continue to debate prevailing models of conception, namely the one- and two-seed theories. A "crucial fundamental problem" remains determining how much significance to allot either men or women or both in conception.[16] Well into the eighteenth century, it goes without saying that women produce semen. This belief is uncontroversial and material to questions of conception and procreation, desire and sexuality, personal hygiene and illness.

The dominant Hippocratic or Galenic stance that female seed corresponds to male seed holds well into the twelfth century. Conception occurs when the twain meet, according to *Secreta mulierum* (*The Secrets of Women*), one of the most widely disseminated gynecological texts of the Middle Ages and early modern period. Composed in the thirteenth century by an unknown German author, more than a hundred copies of the Latin manuscript are made.[17] An early medieval collection of remedies illustrates people's belief that both male and female seed are required for conception. A childless couple can perform a simple (if curious) fertility test to determine where the issue lies. Both are to urinate into separate dishes. By day three, the seed will be visible in the fertile partner's pee.[18]

Many classical ideas regarding the two seeds' influence on offspring carry over to medieval Europe. For instance, a mother's seed may determine her child's sex and conduct certain personal traits. One fifteenth-century medical treatise explains how a baby's sex results from one seed "dominating" the other: if the woman's seed dominates over the man's, the child will be a girl, whereas if the man's seed dominates over the woman's, they will have a son.[19] Noted English philosopher Adelard of Bath (ca. 1070–1152) points out that disease can pass from mother to child, proof positive that the maternal seed plays a role in conception.[20] Konrad von Megenberg (1309–1374), meanwhile, explains in his *Buch von den natürlichen Dingen* (Book of natural things) that the combination of "equally powerful" female and male semen yields hermaphroditic progeny.[21] This is the same explanation we encountered in ancient Indian texts for what were called children of the "third gender."

The Catholic Church recognizes female seed as necessary for conception.[22] Theologian Thomas Sanchez (1550–1610), whose writings define seventeenth- and eighteenth-century Catholic moral theology, views female seed as indispensable for conception, its existence self-evident: "It is most likely that, should the woman fail to secrete seed, impregnation cannot occur."[23] In his foundational text on Catholic moral theology, *Theologia moralis*, Roman Catholic priest and theologian Alphonsus Liguori (1696–1787) subscribes to Sanchez's model of procreation.[24] Not only do these men of God believe that women ejaculate after passionate sex, they also discuss whether men should prolong intercourse until the woman ejaculates. They say yes. Should couples ideally ejaculate at the same time? Also yes. Women should not hold back their seed, not least because it's a sin to refrain from ejaculating.[25] Catholic moral theology expressly forbids both sexes from suppressing ejaculation to prevent conception. *Amplexus reservatus*, or "dry orgasm," in which an individual deliberately contains their efflux, is a damnable transgression for men and women alike.[26] It's not until the eighteenth century that the Catholic Church distances itself from the idea of a female seed.[27]

The Babylonian Talmud, a central text of Judaism, dedicates an entire chapter to female sexual fluids. There is talk here, too, of a yellowish or white discharge: "We can reasonably presume this to be the female prostatic fluid (mixed with other genital fluids of the female ejaculate)."[28] Other works of rabbinical literature also discuss female semen and *zavah*, or female discharge.[29]

In medieval Christian traditions, female sperm is both essential for conception and a dead giveaway for female desire.[30] Women do not release semen any old time, several sources agree, but will instead spray their seed at the height of arousal. The self-proclaimed "discoverer" of the clitoris, Italian anatomist and surgeon Realdo Colombo (1516–1559)—the man behind the term "vagina," as it happens—writes about the clitoral crown as an important center of female desire: "If you not only rub it with your penis, but even touch it with your little finger, the pleasure causes their seed to flow forth in all directions."[31]

The "most ardent champion of female sperm" is encyclopedist William of Conches (died ca. 1150), whose work prompts fierce and varied response.[32] His treatises on natural philosophy, the *Philosophia* and *Dragmaticon*, are widely copied. Conches offers proof that female semen, expressed in a moment of passion, exists. He bases this statement on the supposed infertility of prostitutes: prostitutes don't experience pleasure during sex, ergo no ejaculation, ergo no pregnancy.

Saint Hildegard of Bingen (1098–1179) also writes about female sexuality and pleasure, which she compares "to the radiance of the sun." Medievalist Claude Thomasset notes that the abbess displays some uncertainty when it comes to female sperm. At times "she denies that any such thing exists, at other times she seems to say that women produce small quantities of very weak sperm." She believes that conception occurs when "two frothy liquids (*spuma*)" mix.

The translation of Aristotle—the greatest critic of female seed—raises controversy in the mid-thirteenth century.[33] Thomas Aquinas (ca. 1225–1274) and Albertus Magnus (ca. 1200–1280) are prominent supporters of Aristotle's one-seed model.[34] Although many doctors and philosophers strip the female seed of its generative powers, the notion of female semen is ubiquitous for centuries to come. Another medieval motif is women's enjoying sex more than men. They experience double the pleasure because they both emit semen *and* receive it.[35]

Countless texts warn of the havoc female semen can cause, both in women (as in cases of seminal blockage) and men. Giovanni Savonarola (ca. 1384–1468) explores sex positions in his six-part medical compendium *Practica major*. The Italian physician and scholar states that vaginal penetration in the "natural" missionary position is ideal and adjures men not to allow women on top. His rationale is intriguing: if a woman lies or sits atop her partner, her semen will flow onto him, to his detriment.[36] In *The Perfumed Garden*, Abū ʿAbdallāh Muḥammad an-Nafzāwī more vividly details the dangers of the cowgirl position. This early fifteenth-century Arabic sex manual warns that if a woman mounts her lover, her "seminal fluid might enter the canal of [his] verge and cause a sharp uretritis."[37] Like in old India, female semen is used as a magical elixir in medieval Europe. A medical text from the twelfth century describes the antidote—though not the recipe—to one such love potion: "Aloe helps against bodily blanching and pallor, the result of a man having had female seed slipped in his drink—when a woman desires true love of her paramour, she will have him drink of her seed, from which the man, however, emerges colorless and pale."[38]

The fact that women ejaculate or gush fluids during sex is common knowledge in the Middle Ages. The question is where the fluid comes from. Italian physician Alessandro Benedetti (fifteenth–sixteenth century) describes streams of female semen rising from two passages near the pubic bone, by the opening of the urethra, during coitus. The infertile seed then gushes out with such force that it sprays farther than typical for men. French anatomist Jean Riolan the Younger (1577/1580–1657) associates the fluid with the female prostate. He dedicates an entire chapter of *Schola anatomica novis et raris observationibus illustrata* (1608) to the obscure organ. The "prostates" differ from the male organ in size alone, he explains, and are located near the vagina. They contain a fluid that arouses sexual desire and streams out when a woman is intensely stimulated during coitus. Those who abandon themselves to love, Riolan the Younger contends, will feel this fluid flow near the penis.[39]

Women's health and fertility depend on good seminal hygiene. Sustained abstinence can lead to a life-threatening backup of semen.[40] If a woman fails to ejaculate or gush regularly, her semen can go bad, and, like other bodily fluids, turn into blood or pus, or form an abscess. When a woman is sexually frustrated, her seed stays inside her, potentially blocking the uterus and causing infertility.[41] Spoiled or reserved semen can cause mania or epilepsy, infections, ulcers, discharge, and itching.

Fifteenth-century physicians like Bartholomaeus Montagnana believe that holding back semen is a key cause of hysteria.[42] One late medieval text details a common uterine disorder:

> In cases of suffocation by womb, as this condition is known, a woman will appear close to suffocating as the womb presses against her respiratory organs. This is caused primarily by a backup of monthly flow; poisonous fluid found in the womb; or (the woman's) spoiled, poisonous seminal fluid, which collects around the seminal vesicles in the womb: the latter cause is most common in chaste virgins and widows, in whom, as they abstain from Venusian services, the seminal fluid spoils and turns poisonous.[43]

All it takes is a look, kiss, or light touch to release semen inside the body of a sexually unfulfilled or abstinent woman, causing illness. In the mid-sixteenth century, French physician Jacob Sylvius (1478–1555) paints an impressive picture:

> When a woman, particularly one who is young and vigorous, plump, well nourished, and well supplied with blood and seed, either becomes a nun or chooses of her own free will to remain chastely continent, or marries a man who gives himself to his wife infrequently, or is the widow of a husband much given to this pleasure, or is tempted by the desire of Venus or stimulated by the gaze of some man or by immodest and lascivious talk or by a kiss or by the touching of the nipples or the natural parts as she imagined in dreams, she discharges a large quantity of seed into the womb . . . where it becomes corrupted . . . sending up to the heart and brain certain corrupt vapors that give rise to very cruel accidents.[44]

Orgasm and ejaculation or squirting are thus required to keep women healthy. In married couples, it's the husband's duty to make his wife climax. Women who suffer regular seminal obstruction are advised to take preventive measures, like avoiding foods that increase blood and semen levels, and drinking water over wine.[45] There are countless remedies for relieving the blockage. Wives should simply have more sex (provided they orgasm and ejaculate), while virgins and widows are encouraged to marry. Doctors like Savonarola, wise women, and midwives recommend different methods. The latter, for instance, perform vulval and cervical massages to help ailing women achieve orgasm. Although masturbation is considered a sin in the Middle Ages, it's prescribed as a medical measure against seminal distress. There's any number of medicines, both topical and internal, to trigger the flow of female semen. One such nostrum is an abdominal compress of ground salt stone, sodium bicarbonate, salt water, and vinegar, which causes a biting sensation, and, if the patient is lucky, a spring of seed; failing that, the woman must insert a finger such that its movement and tickling allow the fluid seed to release.[46] Bloodletting, meanwhile, is used to fight prurience (and just about every other medieval malady), but helps not a whit against stuck semen.

The boundaries between male and female bodies are permeable. In one of the tales of *The Decameron*, Giovanni Boccaccio (1313–1375) plays with the changes one's biological body might undergo after that person has shed their social gender. Calandrino, described in the text as a simpleminded painter who appears in four different stories, falls victim to a prank played by his friends, who convince him he's pregnant. He believes them and knows immediately how this could have happened: he allowed his wife, Tessa, on top during coitus.

Tessa having behaved like a man during sex, Calandrino is now pregnant, like a woman.

Another porous boundary—this one between human insides and the world outside—allows for active osmosis. Fluids change freely, one into the other. An English adage of the time illustrates this notion: "Let her weep, she'll piss less."[47] By the same token, men can menstruate or lactate. In the late 1980s, historian Barbara Duden studies one eighteenth-century German doctor's diaristic records and finds that "neither morphological elements nor [bodily] processes like the emission of semen or monthly periods were considered necessarily sex-specific in every instance." According to Duden, blood and milk are not "linked conclusively to the physiological functions of motherhood until the late seventeenth century."[48] Semen will remain a sexual fluid common to men and women for many years to come.

BRIEF ASIDE: FEMALE SEMEN IN THE MEDIEVAL ARAB WORLD

For millennia, it's understood that all humans produce semen, whatever their sex. Medieval Arabic literature and medical texts are filled with accounts of a generative female seed and women gushing fluids in moments of passion. Islamic scholars embrace theories of conception proposed by Hippocrates, Aristotle, and Galen. The notion that both men and women play a part in procreation aligns with Islam. Surah 49 of the Holy Quran reads, "O humanity! Indeed, We created you from a male and a female." Surah 76 states that man was created from a "drop of mingled sperm," and the hadiths explain that women also contribute semen to conception. How else to explain the fact—and we've heard this argument before—that daughters and sons resemble their mother?

Eminent medieval physicians like Avicenna (that is, Abū Alī al-Husain ibn Abd Allāh ibn Sīnā) or Ibn al-Qayyim al-Jawziyya subscribe to the Hippocratic or Galenic two-seed theory. Avicenna (ca. 980–1037), the Persian doctor and philosopher, describes healthy female semen as glistening and whitish, with the scent of palm blossoms and elderberry.[49] There are echoes of Aristotle in Avicenna's characterization of female seed as thinner, weaker, and more blood-like than male semen. The female fluid readies things for the fetus and then empties into the vagina, where it mixes with male ejaculate before being sucked back into the uterus. Experienced at once, these various sensations—the flow of seed, burst of male fluid, and uterine sucking action—are enough to trigger climax in women.

Avicenna is confronted with an obvious conundrum: Why do some women conceive without having orgasmed? How can pregnancy occur without female seed? The doctor riddles it out: whenever a woman becomes pregnant without having ejaculated, it's because fresh male seed has fertilized older female semen still found in the uterus. If a man is fundamentally unable to please his wife in bed, the union will remain childless; a small penis is often to blame for infertility, Avicenna explains, because a woman is less likely to experience pleasure in intercourse, such that she does not ejaculate, and if she does not ejaculate, he concludes, there can be no child.[50] There are serious consequences when a woman doesn't ejaculate. Issues can also arise if she ejaculates, but her semen goes unfertilized and forms abnormal masses in the uterus. These tumors result from masturbation or nocturnal emissions; it is a truth acknowledged by old Arabic physicians and scholars, too, that women experience wet dreams.[51]

Ibn al-Qayyim al-Jawziyya (1292–1351), born in Damascus, believes that male and female seminal fluids are matched in generative power, but differ in other respects. Female semen is yellowish and thin. Rather than spurting, it flows from the body—if at all.[52] There is the notion that, although required for conception, female semen stays in the body. In *The Treatise on Anatomy of Human Body and Interpretation of Philosophers* (1632), Šams ad-dīn al-Itāqī vividly depicts the movement of female semen within the body. The fluid is "yellow like cold mild [*sic*]" and contained in the uterus, where it mixes with male semen—"white and heavy like the rennet"—during coitus: "The male semen has the power of formation and the female semen provides substance. When they are united, harmony appears."[53]

Islamic jurisprudence is influenced by the belief that male and female semen contribute to conception. Legal scholar and theologian Abū Hāmid Muhammad ibn Muhammad al-Ghazālī (1058–1111) argues that coitus interruptus (or *azl*) is necessarily permissible (*mubah*) and contraception does not contravene religious precepts. In one of the "most remarkable documents in the history of birth control," al-Ghazālī reasons that a fetus doesn't form until male and female seeds have mixed in the uterus.[54] Whereas aborting a fetus constitutes a crime against a living being, preventing pregnancy or withdrawing the penis before ejaculation does not. On its own, male semen is nugatory. Any form of contraception that prevents fluids from interacting, whether barrier methods or coitus interruptus, is therefore allowed.[55] Several scholars even sanction (male) masturbation since male semen isn't particularly valuable.

Old Arabic texts do not, however, explain how a woman might curb her seminal emissions to avoid pregnancy. Is there an equivalent to coitus interruptus or *coitus reservatus* in women? *Kitāb al-Ḥāwī* is a monumental medical and pharmaceutical encyclopedia composed in the tenth century by Persian polymath and physician Abu Bakr Muhammad Ibn Zakariya al-Rāzī. The twenty-five-volume work, a detailed survey of knowledge from classical antiquity to the Middle Ages, is one of the most important primary sources on contemporary sexual health practices. *Al-Ḥāwī* alone contains 176 remedies related to conception and abortion, including orally ingested physics, magical tonics, barrier methods (e.g., tampons prepared with honey, pepper, peppermint oil, or dill), such male practices as rubbing the penis in pine tar or balsam oil prior to intercourse, and techniques used primarily by women.[56] Only once, though, does *Al-Ḥāwī* state that pregnancy can be prevented by a man's climaxing first, thus denying his lover her own orgasmic expulsion.[57] Medical figures like Avicenna and Abu al-Hasan al-Tabari adopt al-Rāzī's idea before this method of contraception vanishes from the literature.

Amīn-ad-Daula Abu-'l-Farağ ibn Ya'qūb ibn Isḥāq Ibn al-Quff al-Karaki (1233–1286) authors one of the few Arabic monographs on surgery. In *Basics in the Art of Surgery*, Ibn al-Quff presents his ideas on female semen and uterine anatomy. Following Galen, he interprets female genitalia as male genitalia. Women are colder than men, he states, which is why their reproductive organs are located inside the body, and their seed is thinner and less abundant. Female semen, Ibn al-Quff continues, streams into the uterus from the testicles: "Each 'testicle' (*baiḍatān*) is connected to the uterus by a channel (*mağran*), along which the [female] semen flows. This is known as the seminal sling (*qāḏif al-minā*)."[58]

A more literary take on female seed can be found in *The Perfumed Garden*, a fifteenth-century sex manual by Abū ʿAbdallāh Muḥammad an-Nafzāwī. Like their Indian predecessors, this and other textbooks explain the ins and outs of lovemaking and personal hygiene, but include erotic poems and tales as well. The opening lines of *The Perfumed Garden* celebrate the Creator and physical desire in both men and women:

> Praise be given to God, who has placed man's greatest pleasure in the natural parts of woman, and has destined the natural parts of man to afford the greatest enjoyment to woman.
>
> He has not endowed the parts of woman with any pleasurable or satisfactory feeling until the same have been penetrated by the instrument of the male; and likewise the sexual organs of man know neither rest nor quietness until they have entered those of the female.

The book outlines sexual techniques and problems, discloses sources of pleasure and displeasure, provides treatments for female infertility, and explains how to end a pregnancy. A "meritorious man" will be endowed with a "strong, vigorous and hard" member, "not quick to discharge." Premature ejaculation and quickies are considered unfeeling. Foreplay is key ("toy with her previous to the coition; prepare her for the enjoyment, and neglect nothing to attain that end"), essential to making both lovers' fluids "flow over at the same time."[59]

Old Arabic texts introduce medieval Europe to the significance of foreplay, kissing, and pillow talk.[60] According to Muslim sexology of the time, female desire and orgasm are essential, and ought not to hinge on male orgasm. The idea that a man should tend to female desire, to allow for a woman to reach climax, was missing in Greek and Roman cultures, writes modern-day Islamic theologian Ali Ghandour. In his

2019 study of "Muslims' suppressed erotic heritage," Ghandour quotes a tenth-century text: "Know that the man most loved among women is he who knows how to behave toward them and fulfill their [wishes]." The most important trait a man can have is the ability to recognize female desire and know how to satisfy it.[61]

The Perfumed Garden states that when a man dreams of a woman's vulva—*feurdj*, or "slit"—it means "if he is in trouble God will free him of it; if he is in a perplexity he will soon get out of it," grow wealthy, and settle his debts. Even "more lucky" is to dream of an open vulva. *The Perfumed Garden* describes female genitalia in rich, if ambiguous and uneven, terms: "El maoui (the juicy).—The vagina thus named has one of the four most abominable defects which can affect a vagina; nay, the most repulsive of all, for the too great abundance of secretions detracts from the pleasures of coition. This imperfection grows still worse when the man by preliminary caresses provokes the issue of the moisture. God preserve us from them! Amen." The "biter," meanwhile, is a vulva that "opens and shuts again" on the penis shortly before orgasm. The "large one" is a "vulva which is as wide as it is long . . . , fully developed all round, from side to side, and from the pubis to the perineum. It is the most beautiful to look upon." Here too, seminal emissions are part of female orgasm: "Then do all you can to provoke a simultaneous discharge of the two spermal fluids; herein lies the secret of love." In one story, a "profound connoisseur in love affairs" explains that "in acting thus, the two ejaculations take place simultaneously, and the enjoyment comes to the man and woman at the same moment. Then the man feels the womb grasping his member, which gives to each of them the most exquisite pleasure."[62]

BABY MAKING AND EJACULATING OUR WAY INTO THE EARLY MODERN AGE

The debate over models of conception quiets with the advent of the Renaissance, the period of upheaval in the fifteenth and sixteenth centuries marking the transition from the Middle Ages to the early modern period. In a theory that blends aspects of Hippocratic and Aristotelian thought, doctors and theologians of the era agree that a child is conceived when male semen, female semen, and menstrual blood mix. Nicolas Venette (1633–1698) takes stock of popular assumptions about reproduction and female seed in his best-selling work *Tableau de l'amour conjugal, ou l'Histoire complète de la génération de l'homme* ("done into English by a gentleman" in 1712 and titled *The Mysteries of Conjugal Love Reveal'd*). First published in 1688 and reissued more than thirty times, the book is extremely influential in seventeenth- and eighteenth-century Europe.

Venette—Regius Professor of Anatomy and Surgery and dean of the Royal College of Physicians at Rochelle—lectures extensively on reproductive organs, sexual troubles, and marital dilemmas. He covers the birds and the bees, childbirth, and all manner of illness, and addresses such questions as these: Who experiences keener desire? (Men!) Which season is best for sex? (Spring!) What time of day is most conducive to lovemaking? (Women can do it whenever. For men it really depends on the situation.) What's the ideal sex position? (The professor recommends *a tergo*, "from behind": "This Posture is the most natural and least voluptuous.")[63]

Venette writes an entire chapter on female semen. His observations reveal differences between this and modern-day conceptions of vulval ejaculation or squirting. For instance, in an image we recognize from antiquity, Venette claims that women's seed is produced in the Fallopian tubes and sprays into the uterus more than it does out the body. He believes women play a decisive role in conception: "If Aristotle and his Sectators had not been in great Repute for many Ages past, I am sure it would be easier for me to prove at present that Women have Seed, and contribute a large Share towards Generation. . . . Insomuch that Women have Seed and Terms both together, because they have divers Passions that are evident Signs thereof, the First Matter serving to Engender, and the Second to Nourish the Children they breed." Male and female semen mix in the Fallopian tubes, "and if there be Variety in the Bodies, there is no less in the Souls by the Mixture of the Two Matters and Two Souls."[64]

Venette's theories of embryology are informed by autopsies and other studies, but in particular by his observations of how fertilized chicken eggs develop, which he then applies to human ontogeny. He believes female semen initially feeds the human embryo (as albumen sustains the embryonic chick). Later on, the fetus derives nourishment from menstrual blood, which Venette considers analogous to egg yolks.

Semen also affects women's physical and emotional health. According to Venette, "No Body can deny but that [women] are the most moist of the two [sexes]" and produce more semen, making them prone to sickness and "Fury." They fall ill when their semen spoils, "for when the Seed corrupts, these Distempers ensue," so the professor recommends good, pleasurable sex—and that goes for women too. Men are less susceptible, first because they have less semen, and second because what semen they do have is often released at night, thereby reestablishing equilibrium. In ancient China and India, women were considered superior to men in this regard because their seed was inexhaustible. Venette, meanwhile, believes this to be the very reason women are in such danger. He is certain men experience greater desire than women, pointing out differences in their semen by way of explanation: "Altho' Women are touched to the Quick by the Pleasures of Love when we embrace them; yet I cannot believe that their Sensibility is so great as ours, their Seed being liquid and less hot, is not filled with so many Spirits, and does not sally out with that Swiftness as ours." Venette is familiar with aspects of vulval ejaculation and squirting as we understand it today. For example, he describes the satisfaction (and waning desire) women feel after coming: women "waste by considerable Evacuations."[65]

By this point, it has long been clear that women can get pregnant without orgasming. Nevertheless, common knowledge holds that both lovers' climaxing at the same time provides ideal conditions for conception. The autobiography of Isabella de Moerloose, "wife of a burgher of repute," published in Amsterdam in 1695, reads, "But when both seeds emerge at the same time, a child must necessarily result, because as rennet causes milk to curdle, thus does male seed cause female seed to curdle."[66] In married couples, this notion offers women an excellent opportunity to demand their husbands pleasure them in bed. Childlessness is a sign of an unfulfilling love life, and such couples are plied by doctors and midwives with tips for better sex.

Female seed features in the debate surrounding masturbation, a major focus for philosophers, physicians, and pedagogues starting in the eighteenth century. Swiss doctor Samuel Auguste André David Tissot (1728–1797) is a star player in the antimasturbation league. The methods he recommends in the fight against self-defilement are brutal, ranging from shackles and chastity belts to erection alarm apparatuses, penis sheaths, or clitoral amputation. Tissot's best-selling volume *L'onanisme, dissertations physique et morale*, published in the author's French translation of the original Latin in 1758, was translated into English in 1760 as *Onanism, or, A Treatise upon the Disorders Produced by Masturbation: or, The Dangerous Effects of Secret and Excessive Venery*. In it, the author outlines the horrific consequences of masturbating and squandering "a most important liquor, which may be called the Essential Oil of the animal liquors; or, to speak with more precision, leaves the other humours weak, and in some degree vapid."[67]

Tissot is no quack. One of the most influential physicians of the Enlightenment, he stokes contemporary hysteria surrounding masturbation. He is primarily concerned about the health of young men, but paints a bleak picture of the effects of onanism on girls and women as well. He equates female masturbators with lesbians, and stresses that the outcome of sapphic sex is "equally shocking" and can even lead to death. Women's "humour" being "of more or less value, and not so elaborate as the sperm of men, its loss does not perhaps weaken so soon; but when they are guilty of excesses, their nervous system being weaker than ours, and naturally more subject to spasms, the accidents which arise therefrom are more violent."[68]

In his book, which was an "instant literary sensation throughout Europe," Tissot furnishes one case study after the other to prove the hideous effects of masturbation.[69] When it comes to forced abstinence, Tissot includes a cautionary tale about a woman who,"having been accustomed to frequent emissions, [suspended] them all of a sudden. . . . [She] had been a widow for some time, and the retention of the sperm brought upon her disorders of the uterus; she had in her sleep convulsive motions of the loins, the arms, and legs, which were accompanied with a copious emission of thick sperm, with the sensation as in coition."[70] Interestingly, unlike in the English translation, the 1770 German-language edition describes the woman as experiencing relief, recovering from her condition on ejaculation.[71] Regardless, Tissot wouldn't dream of recommending masturbation to address such uterine affliction, despite his including the example. Even when bad sex or celibacy puts a woman's health at risk, sullying oneself is a far greater threat to well-being than seminal retention.

The antionanism campaign gains greater traction in eighteenth- and nineteenth-century Germany than anywhere else.[72] It's also the age of the great German encyclopedias, composed with equal assiduity. Bookseller and publisher Johann Heinrich Zedler is behind one of these colossal projects. Zedler's *Grosses vollstaendiges Universal-Lexicon aller Wissenschafften und Künste* (Universal lexicon of all sciences and arts) is the most comprehensive German-language encyclopedia of the eighteenth century, containing around 284,000 lemmas in 64 volumes with 4 supplements. Female ejaculation makes an appearance. The lemma on female "lechery" states that women secrete "with utmost pleasure a *serum* known as *Liquor prostatarum*," produced "no differently" than in men, namely by the "*prostatis*, which in women sits on the urethra."[73] Women also experience "pollution," the uncontrolled release of their fluids, and "lecherous thoughts" are known to hound them at the sight of a "pleasant and dear person." If they "itch" their privates—a euphemism for masturbation—a "certain juice" will flow from them. In addition to *Liquor prostatarum*, women produce a second fluid, known as *Liquor genitalis* and found in the "birth members." Men must emit their semen and women their *Liquor genitalis* for a mutually gratifying sexual experience. (Although Zedler discusses the practice, he dismisses female circumcision as a method for combating horniness, arguing that it isn't particularly effective; after all, female lust originates in any number of places.)

The *Universal-Lexicon* entry on the vagina describes numerous large ducts around the urethra that send a watery fluid into the uterus—a fluid that flows liberally and pleasurably from the female "birth member" during "Venusian play."[74] The *Universal-Lexicon* makes no mention of female seed, now a thing of the past. The fluid women spray during sex has no generative qualities. In fact, it serves but a single purpose: making women feel good!

WHEN ONE BECOMES TWO

The eighteenth century gives rise to a new take on women: the female body is no longer the less developed, weaker, or colder version of the male body but rather something all of its own. Differences between the sexes once perceived as fluid or gradational are recast as unequivocal otherness attached to male and female anatomies.

Men and women, now fundamentally dissimilar creatures, stand in mutual opposition. Women are now the Other, exhibiting "typically female" abilities and illnesses that yield a host of new sociocultural tasks and assumptions. Women's bodies, from the skeleton to the nervous system—parts that were always considered universally human—are now "differentiated so as to correspond to the cultural male and female. . . . Two sexes, in other words, were invented as a new foundation for gender."[75]

There's scarcely a part of the female body that doesn't differ from the male, German physician Jakob Fidelis Ackermann (1765–1815) asserts in his dissertation titled "Ueber die körperliche Verschiedenheit des Mannes vom Weibe außer den Geschlechtstheilen" (On the bodily difference between man and woman, other than the genital parts). Ackermann's colleague, renowned gynecologist Dietrich Wilhelm Heinrich Busch (1788–1858), also uncovers gender-specific disparity at every turn: significant differences exist between men and women, be it in the spine, ribs, sternum, cartilage of the epigastric fossa, hip bones, oral cavity, throat, larynx, hand bones, or fingernails ("less dense, more delicate and translucent").[76] Body parts even *smell* typically male or female now.

As autopsies become more common, anatomists dive deeper into the human body. This postmortem practice is the most valuable learning tool available to the fields of biology and medicine. It's a source of inspiration for scholars, who revise their attitudes based on the anatomical and physiological discoveries of the day. The "magical body"—a designation held for centuries by the female body in particular—slowly loses its magic. Male scientists' own interests and ideologies shape new concepts of the female body and "being." Women become "symbolic of a type that can be discovered, deciphered, and examined by the light of reason."[77] The difference between men and women, and the inferiority of the latter, is now traced back to female nature and biology.

Medicine furnishes the arguments for fixing women's societal and family roles. Women bleed once a month? Anyone that helpless belongs in the security of the home, not out in public. Women's brains are smaller and weigh less? Ample evidence that women aren't as clever as men, and that educating girls and women is pretty pointless. This "neutral" proof of women's lesser intellect can be used to "eliminate even the most fervent advocates of female emancipation."[78] Interpretations of the human body run the gamut and are often quite absurd. Noted obstetrician Johann Christian Gottfried Jörg (1779–

1856) writes, "The man prepares his sperm without the woman's involvement and is therefore less dependent on the same in the world."[79]

Primary sex organs now serve as justification for new societal roles: the penis and testicles are found on the outside of the body, "an indication not only of [men's] state of greater completion, but indeed of the overabundant quality of this state." The state of lesser completion, as it were, of female genitals is thus a sign of the inchoate, lacking condition of womanhood. In their seminal work *Die Ehe aus dem Gesichtspunkte der Natur, der Moral und der Kirche betrachtet* (Marriage as viewed from the perspectives of nature, morality, and the church), Jörg and Pastor Heinrich Gottlieb Tzschirner of Leipzig write,

> The incomplete [nature of women] is only too clear in the external composition of the sexual organs, in that these display well enough a partial or unfinished quality. The dependence [on men] of female genital capability is irrefutable in light of the fact that women cannot become pregnant, bear children, or nurse without men. Men, by contrast, appear in a much higher social position and are therefore more complete and less dependent than women.[80]

The authors fail to address the obvious point that men can't grow, bear, or nurse children at all. Some second-wave feminists will later invert the argument, citing women's exclusive ability to generate life as proof of their superiority over men and their less able bodies.

Austrian philosopher Helene von Druskowitz, the second-ever woman to receive a doctorate in philosophy, also demonstrates how the male body can be seen and interpreted in a whole new light. Druskowitz, though committed to a sanatorium for the final twenty-seven years of her life, continues to write and in 1905 publishes *Pessimistische Kardinalsätze* (Pessimistic cardinal truths): "Man is a mediate member between human and beast, a monstrosity so cynically and ludicrously equipped as to be neither fully one nor the other. By means of excessive, conspicuous development, nature has branded his genitals, a routing without equal."[81]

Doctors like Jörg are now the experts on human nature, philosophers of the human body with a penchant for studying and (re)interpreting women's reproductive organs. Periods and pregnancy are now defined as illnesses that bar women from assuming an active, public role in society. Women are the weak, sickly sex whose purview is the home and family life: "Nature has designated the female sex for conception, gestation of the fruit, for bearing and further nourishing the fruit. . . . Nature has therefore allotted women a narrow field," writes Eduard Caspar Jacob von Siebold, director of the "birthing institution" in Marburg, Germany, in the 1820s. The point he's making is clear: "the house is hers; the world, his."[82] In *Dictionnaire philosophique*, even Voltaire (1694–1778) avers that the enigmatic capabilities of the female body, once celebrated, are now the very reason for women's societal subordination. Physiology, menstruation, and pregnancy as well as birthing, nursing, and raising children—all are reason enough to disqualify women from professions that require such unfamiliar things as strength or endurance.

In the late eighteenth and into the nineteenth century, doctors reject many terms that have long been used for body parts and organs. Ovaries and testicles, which once shared a name, are now linguistically distinguished. "Sometime in the eighteenth century, sex as we know it was invented."[83] The male body becomes the standard for all humans, while the female form is a deviation. In late eighteenth-century Germany, this logically gives rise to the "Science of Women," a kind of gynocentric anthropology.[84] People's understanding of female semen changes in keeping with new attitudes toward bodies. The greater the differentiation between genders, the less acceptable the notion of a shared seminal fluid. As the human form splits in two, these bodies' fluids also demand differentiation.

Topics relating to female semen or comparable sexual fluids in women, once standard fare for debate, appear far less frequently in nineteenth-century medical literature. Returning for a moment to Busch, the gynecologist does mention a pleasurable, slick secretion produced during sex in his multivolume *Geschlechtsleben des Weibes* (Sexual life of women, 1839), but the doctor's guarded language exudes uncertainty. Busch also suggests that ejaculation is more common in women who are rather too lusty: "One cannot, with any degree of certainty, disclaim that the female emits a peculiar secretion from the genitals during sexual intercourse, as lecherous women perceive the phenomenon themselves and discharge a small amount of mucus when the sex drive is greatly excited."[85] Otherwise, the doctor only mentions "peculiar secretions" and puzzling mucous effluence when discussing such disorders as hysteria. And thus begins the pathologizing of female sexual fluids.

A key force behind the suppression of female juices is a newfound understanding of human reproduction. In 1672, Regnier de Graaf discovers ovarian follicles inside the ovaries. Around the same time, Antoni van Leeuwenhoeck invents the microscope, and in 1677, he and Johan van Ham see the human sperm cell for the first time. There within the spermatozoon, van Leeuwenhoeck espies a fully developed, tiny human, a cowering homunculus just waiting to be deposited in the womb. In his experiments on dogs about a hundred years later, Lazzaro Spallanzani will prove that the coupling of egg and sperm cells is required for generating new life. In 1875, Otto Hertwig observes a spermatozoon fertilize a sea urchin egg under the microscope. Until around 1900, though, consensus remains elusive on how fertilization actually works, with plenty of competing theories on the matter.[86]

One thing's for certain: for thousands of years, female semen was regarded as equivalent to male semen, albeit often a less perfectly realized, thinner, colder, or weaker version. When the egg cell is identified as women's contribution to conception, however, the sexual fluid once believed to be female seed is deemed unnecessary and superfluous. It's obvious enough why the vagina moistens during sex; penetration wouldn't feel great for either party without it. But if sex is all about reproduction, what's the use of a second fluid? A fluid that doesn't even have a name anymore? "What is known as female seed is nothing more than a mucous substance . . . that provides the vagina with moisture and slipperiness," according to physiologist and pathologist Samuel Schaarschmidt (1709–1747).[87]

Female semen is stripped first of germ, then of term, and that which goes unnamed frequently goes unnoticed. Over time, it changes into everyday mucus, lubrication, nocturnal pollution, or diseased secretion. It becomes the repulsive discharge of wanton women. By the mid-nineteenth century at the latest, semen has become an exclusively male substance; the word "semen" is only ever used in the male context now, and in some cases, like the 1885 *Handbuch der Frauenkrankheiten* (Handbook of female disease), it's referred to by a term that has always been reserved for men alone: sperm.[88]

From time immemorial, sexual drive and pleasure were ungendered facts of life, and it went without saying that both men and women would orgasm during sex. Even after the ova were discovered, female pleasure was considered important for conception. Into the 1840s, many believe that ovulation was triggered by orgasm—in other words, that generative substances weren't produced until folks got busy.[89] By the end of the nineteenth century, this notion is largely superseded by widespread acknowledgment that women can get pregnant without experiencing sexual excitement. Medical discourse explores cases in which women are impregnated in their sleep, while under anesthesia, or as the result of artificial insemination or assault. Heinrich von Kleist's widow in *Marquise of O* (1808) is raped while unconscious and "inexplicably [finds] herself in a certain condition."[90] In *Elements of Hygiene* (1864), the widely read Italian physician Paolo Mantegazza recounts one doctor's horrific attempt to treat a woman suffering from vaginismus, a condition in which vaginal muscles involuntarily seize up or spasm during penetration. The doctor administers chloroform to anesthetize the newlywed and then hands her over to her husband, now free to consummate the marriage. The woman isn't cured, and the couple continues to struggle with intercourse while conscious, but the experiment is deemed a success because it results in pregnancy.[91]

Women are now fertilized by men. Their orgasms are superfluous, "accidental, expendable, a contingent bonus of the reproductive act."[92] Female desire itself is up for debate. Why does it exist, and what's its function? Do women even experience physical arousal and passion? Physicians, philosophers, and writers cook up a new theory: honorable women have a decidedly weak sex drive, if any at all. In the mid-nineteenth century, influential English urologist William Acton (1813/14–1875) declares that desire and lust are alien concepts to decent, respectable ladies: "The majority of women are not much troubled by sexual feelings of any kind."[93] As scientists start delving into female passionlessness and examining the "absence of sensuality in women during coitus," and the first psychoanalysts take an interest in understanding "frigidity" in women, there's a rise in pornographic literature championing women's concupiscence—their desire, bodies, passion, orgasms, and yes, juices.[94]

BRIEF ASIDE: LIBIDINOUS LIBATIONS—PORNOGRAPHIC LITERATURE

> Wenches that are not rich enough to buy statues must content themselves with dildoes made of Velvet, or blown in glass, Frick fashion, which they fill with lukewarm milk, and tickle themselves therewith, as with a true Prick, squirting the milk up their bodies when they are ready to spend.
> —Anonymous, *The School of Venus*

L'École des filles ou La philosophie des Dames is the "most famous and most influential of the erotic books of the period."[95] Published anonymously in France in 1655 and meant for a mixed readership, the novel is translated into English within a year, reaching a huge audience and stirring up lots of attention. In a letter from 1687, French writer Roger de Bussy-Rabutin reports that the ladies-in-waiting of the heir apparent's wife were all dismissed after the book was discovered among their things.[96] Even famed English diarist Samuel Pepys, after reading the "mighty lewd" book in bed, burns it, "that it might not be among my books to my shame."[97] The suspected author of "this first great porn 'novel,'" Jean l'Ange, is condemned to toil as a galley slave—a sentence later reduced to three-years' banishment from Paris.[98]

Female sexuality and desire are at the heart of *The School of Venus*, which depicts wild, fulfilling sex in blunt terms. As one of the protagonists muses, fucking is as natural as eating and drinking. The two-part work, which "does not present a particularly daring sexuality or one that is remote from the practices of the day," follows young Fanchon's *éducation sexuelle.*[99] Still a virgin in part 1, Fanchon learns all about sex from her cousin Susanne. What might her first time be like? The young man "throws her backwards, flings up her Coats and Smock, lets fall his Breeches, opens her Legs, and thrusts his Tarse into her Cunt (which is the place through which she Pisseth) lustily therein, Rubbing it, which is the greatest pleasure imaginable."[100] In part 2, Fanchon and Susanne chat about sex, Fanchon having long since lost her virginity and picked up some experience. Topics range from sexual techniques as well as the length and girth of their lovers' cocks, to the women's experimentation with dirty talk, dildos, contraception, orgasming, masturbation, and public fornication.

The School of Venus is a "libertine gem."[101] The novel shows readers how to get the most out of screwing, illustrating sensual foreplay, exhilarating sex, climax, and orgasmic expulsions. As the man and woman approach orgasm,

> what with rubbing and shuffing on both sides their members begin to Itch and Tickle; at last the seed comes through certain straight passages, which makes them shake their Arses faster, and the pleasure comes more and more upon them, at last the seed comes with that delight unto them, that it puts them in a Trance. The seed of the man is of a thick white clammy substance like suet, that of a Woman thinner and of a red color, mark, a woman may spend twice or thrice to a mans once, if he be any time long at it, some Women have an art of holding the Tops of their Cunts, that they can let fly when they please, and will stay till the man spends, which is a Vast satisfaction to them both.[102]

Five years after *The School of Venus* is published, another novel appears on the scene: *Académie des Dames*, whose title clearly alludes to the earlier work. *Académie des Dames* flies off shelves as well, in multiple editions and translations. In this work too, female seed "decays" if held too long in a woman's loins. It must therefore be evacuated by sexual means. Octavia, one of the main characters, enjoys wonderfully wet orgasms with her husband: "*Venus* was kind unto me, and had, it seems, prepared a Liquor, which flowed from me in such Abundance, as *Philander* himself felt it, and cried Oh! What Joys do effect us both, with such a Stock of this heavenly Nectar, as we are each of us provided with!"[103]

Explosive female fluids are as integral to the pornographic and erotic literature of the time as the ohs! and aahs! accompanying lovers' orgasms.[104] The genre is positively flooded with effluence. Through the late nineteenth century, "to spend" is the most common English term for coming. It applies to male and female orgasm, and includes diffuse fluids on both counts.[105] In his eleven-volume erotic classic *My Secret Life*, the pseudonymous memoirist "Walter" recalls one tryst so charged that "in too short a time we spent together."[106] And the same Octavia quoted above describes her husband "putting his Finger into my C—t, and stirring gently up and down, towards the upper Part of it, [until] he made me spend so pleasantly, such a Quantity of the delicious Nectar, that it flew about his Hand, and all wetted him."[107]

Pornographic literature needn't realistically depict sexuality. Indeed, its existence has often relied on its departure from truth, which makes it hard to know how to interpret instances of expressive female juices in old porn. Are these scenes realistic representations of superannuated sexual practices, the product of an overactive imagination, or an expression of profound misogyny? It's worth noting that female semen features in these works through the eighteenth and nineteenth centuries, despite growing skepticism among medical authorities. Porn has always had a soft spot for the taboo. Are these writers, in depicting female ecstasy in such wet and wild detail, perhaps doing so in response to public suppression of women's pleasure?

Here's to the ladies of lewd literature making a splash before their fluids (almost) entirely dry up. This selection of erotic texts features female juices, and includes such classics as *Fanny Hill* (1749), *My Secret Life* (1890), and *Josefine Mutzenbacher* (1906):[108]

> Our ardours, like our love, knew no remission; and all the tide serving my lover, lavish of his stores, and pleasure-milked, he over-flowed me once more from the fulness of his oval reservoirs of the genial emulsion: whilst, on my side, a convulsive grasp, in the instant of my giving down the liquid contribution, rendered me sweetly subservient at once to the increase of joy, and to its effusions. (*Fanny Hill*)

> "What a lot of spending you have done," said I. "I can't help it," said she. My experience was small, but I knew that from no other woman whom I had stroked, had such an effusion taken place. Before I had spent I felt her wetness on my fingers. I had her on another occasion, and the same thing occurred. (*My Secret Life*)

> Maybe for half an hour he worked me, and I swam in my own juice and bliss. (*Josefine Mutzenbacher*)

4

EJACULATION AND POLLUTION, WET DREAMS, AND JOY FLOW

> The study of women has been paltry and poor. We have complete monographs on the silkworm, June bug, and cat, yet we have none on women. Whence comes this striking wonder?
> —Paolo Mantegazza, *The Physiology of Woman*

If *semen* is considered all but exclusively the preserve of men by the late nineteenth century, this presents a quandary: What does that, then, make the substance known long before as *female semen*? What is this fluid that some women emit during sex, that moistens doctors' probing hands and catches partners off guard? Over the next seventy years or so, doctors and scientists will puzzle over the question, but initially the effluence is grouped with female ejaculation or classified unflatteringly as a pollutant emitted while a woman sleeps.[1]

Among these researchers is Italian physician, philosopher, and early sexologist Paolo Mantegazza (1831–1912), whom we encountered last chapter. Mantegazza's many novels, medical works, and popular science publications are read across Europe and North America, securing his position as one of Europe's most successful writers between 1870 and 1930.[2] Women are superior to men when it comes to sexual potency—something Mantegazza does, however, present as inversely related to intellectual capacity. Not only are women more passionate, he writes in *The Physiology of Pleasure* (1854), his first sexological tract, they ejaculate more frequently, with many experiencing multiple emissions in the same period in which a man will execute just one.[3]

Mantegazza also addresses irksome nighttime pollutions, a major talking point around the turn of the century. In *The Hygiene of Love* (1864), he affirms that men and women both suffer these troubles, though men are afflicted with greater frequency. Girls and women demonstrate surprising variance in the amount of fluid they ejaculate; for instance, voluptuous women (a rarity as it is) have wet dreams most often accompanied by a piddling amount of vaginal liquid. On the other hand, if a hypersexual young woman abstains from sex for a few days, her pollutions will soak the sheets.[4]

Women are confounded by these fluids and turn to doctors for help. Many physicians witness female effluence firsthand during examinations and treatments. Neurologists, gynecologists, psychoanalysts, and sexologists are particularly intent on interpreting the emissions. There's less popular discourse on female ejaculation and pollution than on onanism, impotence, or frigidity, but a range of works do exist—widely read medical publications and pop science books, many of them successful internationally—that fervently debate female sexual fluids. The key voices in this discussion are introduced in the following section.

HEALTHY EJACULATION, SICK POLLUTION

Medical and sexological works of the time distinguish between female ejaculation and pollution. The former is a healthy aspect of sexual climax, whereas the latter is something women and girls experience (or suffer) passively, an involuntary emission of fluid while they sleep, doze, or fantasize about sex. (The topic of pollution is ubiquitous around 1900, perhaps because it allows for female ejaculation without a woman having to be sexually active—a clever move at a time when a sexual woman was increasingly equated with a morally suspect one.)

Mori(t)z Rosenthal (1833–1889), noted Viennese professor of medicine, dedicates an entire chapter of *Klinik der Nervenkrankheiten* (*A Clinical Treatise on the Diseases of the Nervous System*) to hysteria, that most mystifying of female disorders. The seminal work, first published in 1800, is translated into English (1879), French, Italian, and Russian. Rosenthal expands the clinical picture of hysteria to include nocturnal emissions, having "also observed another pathogenic condition, which I have never seen referred to, viz.: *pollutions in females*." He cites several examples, including the case histories of two patients who discharge a fluid while having "voluptuous dreams." The doctor examines the women and discovers a "mucous fluid upon the external genitals . . . produced by the glands of Bartholin and by the acinous glands surrounding the meatus urinarius."[5] His patients admit to reading "light novels" and "obscene books" before bed, which fits the bill. (Young women are as aroused by reading as they are by racy pictures, suggestive song lyrics, or a man's caresses. So says Doctor J. D. T. de Bienville, anyway, who by 1768 is using the term "nymphomania" to identify and "treat" demonstrative, supposedly uncontrollable female desire.) Rosenthal blames "the flux . . . [on] erotic excitement of the nervous system" and treats one of his patients successfully. Soon after the girl's pollutions disappear, and her hysterical paroxysms follow suit.

Richard von Krafft-Ebing (1840–1902) is among the most renowned psychiatrists and sex researchers of the nineteenth century. During his tenure at the University of Graz, the

Austro-German professor publishes "On Pollutive Occurrences in the Female," an essay on the subject of female pollution and ejaculation.[6] The piece appears in an 1888 issue of *Wiener Medizinische Presse*, a regarded medical weekly. Krafft-Ebing states that orgasm is the high point of coitus for both sexes, and that ejaculation—or at least the "ejaculation feeling"—is part of female climax as well. The phenomenon is linked to the mounting arousal of a "reflexive center in the core of the loins." This center can be activated by mental stimuli that lead to pollutions in men and women alike, so, he concludes, the notion of "pollutions in women" is justified.[7]

Krafft-Ebing describes female ejaculation in detail. An "ejaculation center," comparable to what's found in men, regulates the process. Female ejaculate consists of mucous fluids shed by the ovaries and uterus during orgasmic contractions. A sexually frustrated woman is at risk of illness because sex without climax is considered harmful and can lead to hysteria, among other woes. Ejaculation is part of a healthy sex life, regardless of gender, but it's not the only female emission, Krafft-Ebing reminds us. The Viennese expert reports on women passively suffering unconscious or semiconscious pollutions, which he deems a symptom or even the cause of neurosis. He, too, cites myriad case studies to prove his point. He presents one thirty-year-old woman's litany of ailments, including sexual insensitivity, poor sleep, weak and painful periods, "mortifying genital orgasms," and wet dreams.[8] The doctor prescribes his patient sleeping powder, sitz baths, and suppositories of monobromated camphor and deadly nightshade. Krafft-Ebing doesn't say whether the treatment helps dry out her dreams. He was unfortunately unable to conduct any longer observations.

Krafft-Ebing also discusses female effluence in his foundational reference work *Psychopathia Sexualis* (1886). He warrants that "[in] the female the pleasurable feeling occurs later and comes on more slowly and generally outlasts the act of ejaculation." From the outset, Krafft-Ebing establishes that "man has a much more intense sexual appetite than woman," and if healthy and "well bred," a woman's "sexual desire is small."[9] Still, for both parties, "the distinctive event in coitus is ejaculation."[10]

Krafft-Ebing analyzes the minutiae of human sexuality and investigates other areas, like the relationship between psychiatry and criminal law. In 1886, he points to the shared qualities of male and female ejaculation in a novel potshot at paragraph 175 of the German penal code, which criminalizes "unnatural sexual acts" between men. Given that both sexes ejaculate, he argues, the rule could also be applied to women. Women are equally capable of resorting to "unnatural means of sexual satisfaction"—that is, engaging in sexual behavior with other women—because given enough stimulation of her erogenous zones, "even a woman" will experience something analogous to male ejaculation, meaning the act leading to that point must be equivalent to coitus.[11] Krafft-Ebing objects to what he sees as the statute's inconsistency, naivete, and flawed logic. Paragraph 129 of Austrian law, meanwhile, "takes cognizance of unnatural sexual relations . . . between those of the same sex," which Krafft-Ebing figures could therefore be applied to female couples as well.

Berlin-based research physician and sexologist Albert Siegfried Jakob Eulenburg (1840–1917), a contemporary of Krafft-Ebing's, studies female ejaculation and pollutions too.[12] As it relates to healthy female orgasm, Eulenburg describes the "erection of the clitoris and ejaculation of a mixed secretion from the Bartholin's glands, cervical mucous glands, etc."[13] He writes that ejaculation is reflexive and triggers a shift in the uterus critical for impregnation—a theory that reestablishes a causal link between orgasm and conception.

Unlike ejaculation, which is healthy and normal, female pollutions are deemed "vulvo-vaginal crises" or "clitoris-crises." Eulenburg classifies three types of pollutions: pathological pollution, masturbatory emissions (Eulenburg opposes onanism, and warns parents and teachers against making light of young girls' enthusiasm for it), and nocturnal events in hysterical women. This final class, with its "peculiar vulvo-vaginal emanations," is the "most manifold and interesting form of secretionary disturbance." Some hysterics, Eulenburg quotes his French counterpart Fabre, "cry from their vulvas."[14]

Doctors Otto Adler, Albert Moll, Enoch Heinrich Kisch, Wilhelm Stekel, and Max Marcuse are among the loudest voices in the discussion of sexuality and female sexual disorders from 1900 on. They run private practices or work at sanatoriums and health resorts. Women at this time are drawn especially to private consultations with doctors and representatives of the newish field of sexology, like Berlin-based physician Magnus Hirschfeld, who publishes the first *Journal of Sexology* in 1908 and then opens the first Institute of Sexology in 1919. There's a personal quality to the sessions, which creates "a feeling of familiarity" and makes it easier for female clients to discuss intimate topics.[15] It also allows doctors to mine these confessional accounts, tales of suffering, and clinical observations for their publications.

For a few years, medical and sexological discourse returns its focus to the similarities between male and female bodies. Doctors think of genitals and such processes as arousal, pleasure, and climax in terms of analogies, with a pronounced emphasis on the parallels between the sexes.[16] Contemporary texts are riddled with the phrase "as in the male." Given that satisfying intercourse culminates in orgasm for both sexes, ejaculation—the climax of the climax, as it were—is therefore part of the experience for men and women alike. The *Bilderlexikon der Sexualwissenschaft* (Illustrated lexicon of sexology), published in 1930, defines "Ejaculatio" as the "emptying of sexual matter": "The process of e. [ejaculation] in the female is the very same as in the male."[17]

Kisch, a gynecologist and balneologist (an expert in the therapeutic use of baths) based in the spa town of Mariánské Lázně (Marienbad), takes a downright radical tack. In 1904, he publishes *Das Geschlechtsleben des Weibes in physiologischer, pathologischer und hygienischer Beziehung* (*The Sexual Life of Woman in Its Physiological, Pathological and Hygienic Aspects*). As far as Kisch is concerned, female ejaculation is no anomaly. It's to be expected in "the woman whose sensibility is normal."[18] Arousal and ejaculation are so closely related, to his mind, that Kisch uses the phrase "voluptuous sensation of ejaculation" synonymously for orgasm: "Although until recently the matter received but little attention, it must now be regarded as a well-established fact, that in the female (as in the male) the climax of voluptuous sensation in sexual intercourse is normally characterized by a process of ejaculation, accompanied by a voluptuous sensation of ejaculation."[19]

Intercourse that does not lead to a woman's ejaculation or sensation of ejaculation is ungratifying and "deleterious," the health consequences of which Kisch adjures fellow physicians not to underestimate.[20] Libido being one of the most powerful human drives, Kisch argues that its gratification is paramount for women's general health, standing in marriage, and reproductive capacity. He views the lack of ejaculation as a symptom: "The constant sign of dyspareunia [i.e., apathy, discomfort, or the absence of 'voluptuous sensations' during sex] is the failure of ejaculation during coitus." "Incomplete coitus," in which a woman does not ejaculate, leads to "a series of nervous troubles," "great emotional depression," and "at times even . . . melancholia." Kisch also clears up the mystery regarding the fluid's origins: ejaculate is a mixture of secretions from the Bartholin's glands along with uterine and cervical mucus.[21]

The Sexual Life of Woman is a smash. Translated into English, French, and Spanish, by 1917 the original German title is already in its third edition. Anyone tempted to hail Kisch as an early rediscoverer of female ejaculation, however, is terribly mistaken. On the contrary, Kisch's books illustrate how, almost overnight, ejaculation is recast as a suspect, pathological, and rare phenomenon, no longer the self-evident part of female sexuality it once was.

First, though, let's take a look at Adler (1864–?), a Berlin-based gynecologist and prominent figure in the debate surrounding female sexual disturbances.[22] Adler, like so many sexologists at the turn of the century, draws from patient records for his publications. He observes pollutions most frequently in women "who are forced to do without the sensuality they once acquired and exercised. [Pollutions] are the bane of young widows."[23] As for semantics, he objects to the fact that the same word—pollution—is used for a process that presents so differently in men and women. The fluid women produce, Adler says, is neither generative nor can it be described as bursting from the body.

In 1904, Adler writes a fascinating book on women's "absent libido" and "absent orgasm." The work is well received, with the third edition appearing by 1919. In *Die mangelhafte Geschlechtsempfindung des Weibes* (Deficient sexual sensibility in women), Adler states that many women don't enjoy sex and never reach climax. Thank goodness for the new field of sexology and steady flow of publications it's producing: "Out of the sexual darkness and enigmatic stillness comes the 'sexual Enlightenment.'" Adler quotes one L. H. von Guttceit, a doctor, who believes 40 percent of women to be "frigid," without any "sense of the tremendous pleasure of ejaculation." Adler is convinced that "ejaculation, a discharge of fluid, a 'wetting,'" occurs during climax in women and laments his colleagues' ignorance about it. "Libidinous effluxions" rarely "seek consultation," he writes, and women are reluctant to mention such fluids, either because they're unaware or ashamed. Whenever women do confide in their doctor, these genital fluids are usually misperceived, or misdiagnosed as pathological vaginal secretions or "vaginal catarrh."[24]

Adler is also interested in where ejaculate originates, concluding that it comes from "various sources," a mixture of uterine mucus, discharge from the vaginal mucosa, and secretions from vulval vestibular and Bartholin's glands. Fluid emerges at the onset of arousal, and "expression occurs at the height of sensation, upon orgasming, with the contraction of the muscle apparatus considered here." Ejaculation in women is, however, less relevant than in men: "The lesser value of female sexual secretions can easily be explained by their tenuous role in impregnation. Whereas male sperm is the sole purpose of ejaculation, to deposit live spermatozoa in the vagina . . . , the female ovum (equivalent to spermatozoa) . . . has long been waiting in the tube or uterus. At the moment of orgasm, the female organism is not responsible for creating any new objects for fertilization."[25]

Adler provides a number of case studies in his book. Watching a female client masturbate, the doctor observes, "At the height of ecstasy, a light effluence emerges that makes the hand noticeably wet, and whose smell differs clearly from the usual (odorless) vaginal fluid." He also records the account of one heterosexual couple's "simultaneous ejaculation." Regarding another client, a thirty-year-old woman, Adler writes, "The patient was sensible of the man's ejaculation during complete intercourse as a light pressure *in vagina*. Her own effluence occurred simultaneously, dampening the outer genitalia considerably."[26]

Adler's remarks on female ejaculation are ambivalent. On the one hand, the doctor presents it in a positive light as a pleasurable part of female orgasm, while on the other, he interprets ejaculate as an inferior fluid without generative power whose ultimate purpose is to aid in heterosexual, vaginal intercourse: "The function of a woman's ejaculate is largely mechanical, easing *immissio penis* and preventing an overly forceful violation of the vaginal walls by friction. It would be a failure of nature to create a secretion for orgasm alone, when, by the time it is needed, it is already too late!"[27]

Stekel (1868–1940), a Viennese physician, has a unique turn of phrase for nocturnal emissions. Stekel, who authors more than five hundred widely read publications, thus contributing significantly to the popularization of psychoanalysis, views pollution as an act of concealed masturbation. He calls it "larval onanism" or "slumbering onanism."[28] Unlike so many of his contemporaries, Stekel doesn't condemn masturbation; instead, he considers it a form of self-love worthy of medical and societal acknowledgment.

Moll (1862–1939), a physician and psychotherapist, is a founding father of modern sexology, whose studies of childhood sexuality influence Sigmund Freud's work.[29] With a private practice on Berlin's fabled Kurfürstendamm, Moll is a leading authority on sex around 1900, and enjoys high standing among colleagues in the fields of psychotherapy and parapsychology.[30] He writes extensively on female ejaculation, including several case studies in his 1898 *Untersuchungen über die Libido Sexualis* (*Libido Sexualis: Studies in the Psychosexual Laws of Love*, translated into English in 1933). By no means denying the existence of female ejaculation, Moll does—like Adler—devalue it on various counts, always in comparison to the male. Ejaculation does not occur in all women. Female orgasm does not feature ejaculation by necessity. Women express less ejaculate than men. Given that female ejaculate contains no sperm, it is, of course, meaningless with regard to reproduction. Moll suspects the fluid comes from the Bartholin's glands and maybe the "uterine mucous glands." He concludes that female ejaculation doesn't matter for a woman's sexual gratification or "have the same significance as male ejaculation for the purpose of copulation."[31]

The popular perception of ejaculation is gradually changing during this time, as is how to classify it. Instead of a healthy bodily function, female effluence is increasingly treated as a pathological disorder, a tangible expression of deviance or illness. There's a long list of doctors, sexologists, analysts, and gynecologists busy publishing on female ejaculation and pollution, in particular, from the turn of the century into the 1930s. All the more astounding, then, that the topic will have vanished from medical discourse within a few decades. In *Sexual Behavior in the Human Female*, released in 1953, Alfred C. Kinsey states that ejaculation is "the only phenomenon in the physiology of sexual response which is not identically matched in the male and the female, or represented by closely homologous functions." The substance often identified as ejaculate is just "some of the genital secretions."[32]

Kisch's output exemplifies how quickly talk of ejaculation is snuffed out. As a reminder, in 1904 Kisch declared that every "woman whose sensibility is normal" ejaculates.[33] Less than fifteen years later, he writes a new book, another bestseller. *Die sexuelle Untreue der Frau* (Sexual infidelity in women), published in 1917, reads as if written by a different author altogether. Kisch's views on bodies, pleasure, orgasm, and ejaculation have changed radically. He insists on absolute differentiation between male and female anatomy, physiology, and even pleasure, all now diametrically opposed. Men are active; they attack and vanquish. Women, by contrast, are the "waiting, receiving, conceding, passive party." "As in the male" now reads "unlike in the male." Men are compulsive and lustful, as reflected in their uncontrollable erections, the "engorgement of the mighty erectile member." What about ejaculation and erectile tissue in women, though, or the fact that their genital blood supply exceeds that of men? What *about* it? Kisch now defines female ejaculation as the "discharge of certain secretions from genital mucous membranes and a pleasant sensation of secretion," and casts it as a "weak analogy" of male ejaculation. Furthermore, that pleasant sensation is "unknown to most married women their whole life, familiar only to those with erotic knowledge and experience in the secrets of love."[34]

Erotic potency in "normal" women has all but disappeared. Their orgasms are now qualitatively different from men's, far more complicated and nigh impossible to achieve. Effective immediately, ejaculation is reserved for a select few (morally suspect) women. Kisch also eradicates involuntary nocturnal pollutions from his new model of female sexuality: "Very rarely—and only in lascivious, perpetually aroused women—do pollution-like emissions from female genital glands occur in response to erotic dreams."[35] Proudly reflecting on more than fifty years' experience as a gynecologist and boastful of his brimming professional dossier, Kisch suddenly pivots, reinterpreting his observations, research notes, and patient reports after only a few years have passed. His books illustrate the rapid deterioration of the significance of female ejaculation in the first three decades of the twentieth century. They also demonstrate how unfathomable ejaculation becomes when male and female bodies are presented as opposites. Historian Thomas Laqueur writes, "I discovered early on that the erasure of female pleasure from medical accounts of conception took place roughly at the same time as the female body came to be understood no longer as a lesser version of the male's (a one-sex model) but as its incommensurable opposite (a two-sex model)."[36] If women are distinct from men, if they differ in fundamental ways, both physically and mentally, then ejaculation "as in the male" is unimaginable. The gender binary, which presents biologically born men and biologically born women as utterly unlike creatures, doesn't have room for women's sexual power—or their ejaculation.

PROJECTILE PLUGS OR ACTUALLY JUST PEE?

Early twentieth-century doctors are hell-bent on determining the composition of female ejaculate. Most believe it to be a cocktail of fluids drawn from the ovaries, uterus, and vagina, mixed with secretions from the Skene's and Bartholin's glands. Hermann Oscar Rohleder (1866–1934), a "sex doctor" and physician based in Leipzig, thinks female ejaculation is the "expression of uterine and cervical mucus."[37] In 1926,

Hirschfeld writes that female pollutions do not contain gametes but instead are comprised of "cervical plugs," a collection of mucus that fills and seals the cervical canal during pregnancy.[38] Hirschfeld's Austrian colleague Bernhard Bauer (1882–1942) also promotes the theory that cervical mucus plugs are "spasmodically spurted out," and that this substance is female ejaculate: "At the same time, according to various authors' descriptions, a mucus plug—situated in the cervix when the individual is at rest—is then ejected from the same to make way for incoming spermatozoa in the male seminal fluid. In contradistinction to the thin mucus emitted from the aforementioned glands [vaginal mucosa] at the onset of tumescence that coats the mucous membrane, the mucus plug is thick and viscous."[39] According to the gynecologist in his 1923 book, *Wie bist Du, Weib?* (What are you like, woman?), the objective of sexual intercourse is "the ejaculation of seminal fluid in men, and the ejaculation of the mucus plug in women."[40] In other words, women ejaculate to clear the path for fertilizing sperm.

Sexuality starts edging into the emotional sphere in the late nineteenth century, increasingly seen as a "holistic experience involving the brain and nervous system."[41] The mind regulates arousal, pleasure, and orgasm, while the sex organs carry out its orders. The purpose or aim of sex gradually changes as part of this process, from the commingling of both parties' generative substances (first male and female seed, and then sperm and egg cells), with the goal of conception, to subjective gratification and personal sexual release.

In 1900, Freud and his disciples join the debate on female sexuality. Psychoanalysts are more interested in the mental aspects of sex than in bodily concerns. As we've seen, medical and sexological texts from the 1920s and 1930s, which emphasize disparities between male and female forms, are a death knell for common awareness of vulval ejaculation. After all, the idea that women ejaculate conflicts with a strict differentiation between the sexes. Increasingly men are defined as active, hard, penetrative, and prone to erections, whereas women are passive, receptive, and soft; the more broadly these definitions are accepted, the more obvious it seems that ejaculation is a male phenomenon, and ejaculate a male fluid. As Stekel writes in 1920, "*The woman of the future is not a 'she-man'; she is a 'full-woman.'*" Stekel further reasons that "[a] woman can never become a man; the feminine *psyche* is too intricately bound up with the womanly *physis*."[42] That woman of tomorrow shall be a feminine lady, one who receives male semen without any fuss, not one who gushes in pleasure.

The generative female seed of bygone centuries has been reduced to a nigh unknown, unnameable fluid. What, then, is the wetness some women still claim to produce? Where does it come from, and what function does it serve? It isn't required for pregnancy or penetration, plus it only appears in some women (and sporadically, at that), so could it be it's not a discrete substance? Is the liquid some women spritz during sex actually just pee? Is female ejaculation a symptom of stress incontinence—like the leakage some women experience when they sneeze, cough, or laugh—that requires medical attention? In his influential work *Die männlichen und weiblichen Wollust-Organe des Menschen und einiger Säugethiere in anatomisch-physiologischer Beziehung* (The anatomical-physiological relationship of male and female organs of sexual arousal in humans and other mammals), German anatomist Georg Ludwig Kobelt (1804–1857) provides an impressively detailed description of the clitoris (complete with inner structures, nerves, blood vessels, and surrounding musculature) and precise portrayal of female arousal. In the same book, which appears in 1844, Kobelt also mentions the "involuntary micturition" some women experience during intercourse.[43]

It's not until the twentieth century, however, that ejaculation, urine, and supposed bladder weakness are brought into portentous association. Physicians, sexologists, and psychiatrists like Havelock Ellis—who in 1903 asserts that women's accidental peeing during sex is a frequent, normal concomitant—prepare the ground for this interpretation, echoes of which can be heard to this day. In Berlin, Marcuse (1877–1963), a sexologist, equates female ejaculation entirely with incontinence, which he considers the "equivalent of sexual secretion."[44] German doctor and sexologist Ernst Gräfenberg (1881–1957), who publishes groundbreaking texts on the female prostate and anatomy in the 1940s and 1950s, studies this matter with greater assiduity than many of his compatriots. He observes a *femme fontaine* having an orgasm and then examines her ejaculate:

> If there is the opportunity to observe the orgasm of such women, one can see that large quantities of a clear transparent fluid are expelled not from the vulva, but out of the urethra in gushes. At first I thought that the bladder sphincter has become defective by the intensity of the orgasm. . . . In the cases observed by us, the fluid was examined and it had no urinary character. I am inclined to believe that "urine" reported to be expelled during female orgasm is not urine.[45]

Unfortunately, Gräfenberg—whom we'll encounter later as the central rediscoverer of the female prostate and ejaculation—is an exceptional figure among his ilk. William H. Masters and Virginia E. Johnson, famed for their pioneering research into human sexuality in the 1950s and 1960s (and whose story was successfully adapted for television in *Masters of Sex*, which ran on Showtime for four seasons from 2013 on), establish that only a handful of their female test subjects ejaculate. Although they concede that the fluid these women emit is not urine, Masters and Johnson declare that female ejaculation is typically an instance of stress incontinence. Their recommendation? Either toning the pelvic floor muscles or surgery: "Since this condition is usually correctable either by the use of Kegel exercises or minor surgery, medical evaluation is warranted if a woman is bothered by such a response."[46]

This misinterpretation of female ejaculation—the one that claims it's basically bed-wetting—delivers a lasting blow to the sexual phenomenon. Female ejaculation is rapidly suppressed in the twentieth century, and the taboo lingers today. Accidentally peeing at the height of release is a mortifying thought for many women. What a nightmare, to piss on one's partner during sex. In the twentieth century, a natural aspect of female sexuality turns into something shameful that women strain to hold in. Or they try not to climax in the first place. Healthy women are pathologized (FINDINGS: incontinence) or even worse, operated on.[47]

As a side note, this confusion shows up in anthropology as well. In *The Sexual Life of Savages* (1929), for instance, Bronislaw Malinowski writes that Trobriand Island women pee during orgasm.[48] Otto Finsch observes similar behavior among the aboriginal women of Pohnpei, and in *Truk: Man in Paradise* (1929), Thomas Gladwin and Seymour Bernard Sarason report from the South Pacific that "female orgasm is commonly signaled by urination."[49]

In 1982, Joseph G. Bohlen, MD and PhD, reflects on sexological research of the past. Prevailing thought, he writes, was that all female orgasmic fluids were urine, though he also cautions against believing the same all to be ejaculate.[50] Eve Ensler recounts a traumatic experience with ejaculation from the 1950s in *The Vagina Monologues*: "And I got excited, so excited, and, well, there was a flood down there. I couldn't control it. It was like this force of passion, this river of life just flooded out of me, right through my panties. . . . It wasn't pee and it was smelly—well, frankly, I didn't really smell anything at all. . . . I was 'a stinky weird girl,' he said."[51] In *The Clitoral Truth*, women's health advocate Rebecca Chalker relates another woman's tale:

> I've got a long list of things to thank the feminist health movement for, but high on the list is knowing about female ejaculation. During sex, I worried and got embarrassed about the mess I seemed to make. Until I learned about the full anatomy of the clitoris and that other women had actually ejaculated, I held back during sex (with both women and men)—I had learned my lessons about wetting the bed. . . . I don't always ejaculate every time I make love, but now I know that I don't have to hold back. That's the main message to take from learning about female ejaculation: don't hold back, and at the risk of sounding clichéd, let it flow![52]

To this day, it's hard to find a report on vulval ejaculation and squirting that doesn't differentiate between the fluids and urine—or conversely, doesn't insist squirting *is* urine, little more than stress incontinence.[53] A British study from 2007 encourages women to enjoy ejaculation, provided they've ruled out urological indications for the spray.[54] In a study conducted by Swedish researcher of sexology Dr. Jessica Påfs in 2021, nearly all twenty-eight participants voice uncertainty regarding the makeup of the fluid they ejaculate or squirt. Påfs writes, "Almost all women had either smelled, or tasted, it to confirm that the expelled fluid was not urine and concluded it was different."[55] These women's experience of coming runs the gamut. To some it's incredible, like a physical superpower or feminist statement. Others feel discomfort or shame, the primary reason for which appears to be fear that the fluid may contain urine.

But let's get back to the twentieth century and the question of where the juice known as "female ejaculate" comes from. What organ might be responsible for this contentious substance? In men, 65–70 percent of ejaculate comes from the paired seminal vesicles (*glandula vesiculosa*) located near the prostate, which itself contributes 10–30 percent of the fluid. The testicles and epididymides generate about 5 percent, while the paired, pea-sized bulbo-urethral glands add 2–5 percent of the overall composition. The average orgasm will produce about one teaspoon of semen. In men, too, the amount, smell, color, taste, and consistency of ejaculate changes, depending on things like age, diet, physical activity, stress, and temperature. When men come twice in a row, they produce less ejaculate the second time round—unlike women, incidentally. The female homologue to the bulbo-urethral glands is a pair of large vestibular glands (*glandula vestibularis major*) known as the Bartholin's glands, which empty into the vulval vestibule. Cis women's ejaculate emerges from the urethra or vagina, though, so the fluid can't originate in these glands. Bulbo-urethral glands, testicles, and epididymides are only found in men, which rules them out. That leaves the prostate, which produces a great deal of male sexual fluids. But does a female prostate exist?

BRIEF ASIDE: THE FEMALE PROSTATE, PART I

Society has always shaped the study of the human body, but cultural prejudice, desire, and ideology have an outsize influence when it comes to learning about sexual and reproductive organs.[56] Time and again, the female prostate has fallen victim to bias. This organ has been studied and written about for hundreds of years. Its history—like that of the clitoris or ejaculation—is one of discovery, description, and disregard tailed by rediscovery, revised description, and renewed disregard. Given that ejaculate is largely produced by the *Prostata feminina*, a history of female ejaculation is at once a history of the female prostate.

The male prostate, similar in size and shape to a walnut, is located near the bladder and surrounds the urethra. In women, the gland's positioning and shape run the gamut. The female prostate may be located at different spots along the urethra and range in form. Between individuals, the number of ducts in the prostatic tissue may also vary. This distinctive feature complicates the study of the organ, yet explains why some women don't ejaculate.

Although Herophilus (ca. 325–255 BCE) and Galen wrote about "glandular helpers" situated on either side of the bladder or around the bladder neck, and the role these body parts played in producing male and female seed, the prostate was not recognized as an organ until the Renaissance.[57] In *A Cultural History of the Prostate*, Ericka Johnson, a professor of gender and society at Linköping University in Sweden, writes that "by the 1600s, the existence of a prostate (or two) was part of the standard core of knowledge about men's reproductive tracts; the gland had been drawn, and given a medical name: prostatae."[58] The female prostate appears in the writings of Italian physician Arcangelo Piccolomini (1526–1605), and Jean d'Astruc (1684–1766), professor of medicine in Paris and Montpellier, mentions the gland in a treatise on venereal diseases. He identifies it as homologous to the male prostate and notes that it surrounds the woman's urethra. In his book on gynecological disorders, British obstetrician William Smellie (1697–1763) discusses a thin fluid expelled during sex and produced by two prostatic vesicles that empty into the urethra. Some women emit up to 1–2.5 ounces, Smellie observes.[59]

In 1853, Rudolf Virchow (1821–1902) mentions the "paraurethral" glands (prostatic tissue around the urethra), which he views as homologous to the male prostate. In the late 1800s, doctors G. Oberdieck and Ludwig Aschoff describe prostate-like structures along the urethra.[60] Gustaf Pallin identifies the female prostate as homologous to the male gland in 1901, as does Walther Felix in 1911.[61] US anatomist Franklin P. Johnson creates a wax model of the prostate in 1922. Russian gynecologists E. N. Petrowa, C. S. Karaewa, and A. E. Berkowskaja study the serial sections of sixty female urethras in the 1930s, publishing their findings widely, including in Germany.[62] In 57 percent of the samples, the team discovers clearly articulated paraurethral glands. In several cases, one part of the glands is found to display full morphological similarity to the prostate.[63]

In 1941, pathologist George T. Caldwell examines 105 female urethras. He, too, discovers paraurethral glands that often contain fluid and differ significantly in size, regardless of the woman's age. In a study published five years later, Caldwell and coauthors Russell L. Deter and A. I. Folsom classify 100 female urethras; 8 percent do not appear to have paraurethral glands. In 18 percent, some articulation is evident. Moderate articulation of glandular tissue is found in 29 percent, while in 28 percent of those studied, greater articulation is found. The entire urethra is surrounded by prostatic tissue in 17 percent of the urethras examined. The authors conclude that "the name 'female prostate gland' could be used instead of 'female peri-urethral glands' in order to emphasize the homology of these glands in the male and female."[64]

The biggest names on this roster of medical discoverers are Dutch anatomist and physiologist Regnier de Graaf (1641–1673), twentieth-century gynecologist John W. Huffman, who studies the female prostate extensively, and Gräfenberg.[65]

De Graaf is the first to publish a detailed description of the clitoris, female prostate, and prostatic ducts.[66] In a 1672 paper on human reproductive organs, he presents the prostate in women as an articulated glandular structure surrounding the urethra:

> The urethra is lined internally by a thin membrane. In the lower part, near the outlet of the urinary passage, this membrane is pierced by large ducts, or lacunae, through which pituito-serous matter occasionally discharges in considerable quantities.
>
> Between this very thin membrane and the fleshy fibres we have just described, there is, along the whole duct of the urethra, a whitish, membranous substance about one finger-breadth thick which completely surrounds the urethral canal.

According to the Dutchman, this substance could "be called quite aptly the female prostate or corpus glandulosum, 'glandulous body.'"[67] De Graaf is also convinced of the prostate's purpose being to "generate a pituito-serous juice which makes women more libidinous with its pungency and saltiness." He notes that the discharge from the female "prostatae" causes as much pleasure as does that from the male "prostatae," and could therefore be described as female pollution. Although the substance is not real semen, de Graaf writes, a number of women admitted to acting as reprehensibly as men who defiled themselves; while entertaining "lascivious thoughts," these women (sinfully, and counter to any sense of decorum) moved their fingers back and forth or employed "suitable instruments" until they reached climax, whereupon they ejaculated liberally. He states that some of the "pituito-serous substance" undoubtedly came from the nerve-laden vaginal membrane or ducts within the same, but that anyone who examined the female prostate's dendritic passageways about the urethra would find that most of the fluid was emitted here.[68]

How could we lose this precise understanding of the female prostate—much of which aligns with the state of knowledge today—in the intervening two hundred years? Two anatomical discoveries are largely responsible for the organ's disappearance: Caspar Bartholin's discovery of the greater vestibular glands (*Glandulae vestibulares majores*, hereafter known as the Bartholin's glands) and Alexander Skene's discovery of glands that become known as the Skene's glands.[69]

Not long after de Graaf publishes his findings, Bartholin (1655–1738), an influential Danish physician, dissects a single (!) female cadaver, and instead of a prostate, finds two glands that empty into the urethra. These glands, however, are far too small to produce much fluid. Bartholin therefore concludes that what he has discovered is responsible for ejaculation, not de Graaf's prostate. He proposes that the glands and their ducts—described as "more open," and located near the urethral and vaginal openings—are suitable for excreting the fluid. This is especially likely, Bartholin explains, because on examination, they seem to originate from glandular tissue similar to the male prostate gland, or at least similar to the tissue that ran through the urethral ducts described by de Graaf.[70]

Bartholin is so imprecise in describing the location of these glands that they are subsequently confused with the female prostate.[71] To wit, in 1886, Eugène Guibout writes that during coitus, secretions from the Bartholin's glands will suddenly increase. He describes voluptuous women, whose keen appreciation for the sensations of procreation induces a veritable flooding of the genitalia. He acknowledges that—in all the excitement—other mucous glandular vesicles in the vagina and vulva supply their own damp, slick stuff, but maintains that it's the Bartholin's glands that contribute the most. In some extreme cases, he observes, the secretion will shoot forcefully from the glands. Though discussing female ejaculation, Guibout links it to the wrong glands. In the 1879 *Handbuch der allgemeinen und speciellen Chirurgie* (Manual of general and special surgery), author Adolph Wernher traces ejaculate back to the Bartholin's glands. Based on his understanding of the "female sexual apparatus" and its various functions, he concludes that "these ejaculations" must come from the Bartholin's glands. This conflation is fatal. The glands Bartholin discovered are very different from the female prostate: they're much smaller and produce barely any fluid. After Bartholin's study, de Graaf's description of the female prostate is gradually forgotten, as Austrian physician and ejaculation researcher Karl F. Stifter writes.[72] The Dane's work has led to confusion that lingers today.

Alexander Skene (1838–1900) discovers another pair of glands, even smaller than the Bartholin's glands, yet these too pose an existential threat to the perception of the female prostate and ejaculation. In 1880, the Scottish gynecologist describes "two important glands," two "tubules [that] run parallel with the long axes of the urethra," and empty into the vulval vestibule and lower section of the urethra.[73] The "Skene's ducts" join the list of parts subsequently mistaken for the female prostate, like in Theodor Hendrik van de Velde's 1926 classic *Die vollkommene Ehe* (*Ideal Marriage: Its Physiology and Technique*). Over time, the prostate diminishes as the complex structures de Graaf described are forgotten. "Illustrations in modern anatomy books," writes physician Sabine zur Nieden in 1994, "borrow heavily from Skene's graphics, despite later authors' having proven that this glandular structure is far more articulated and varied in form than the paraurethral glands' two primary ducts, as described by Skene."[74]

In the late 1940s, Huffman, a gynecologist, creates a vivid, "enduringly revolutionary" image of the female prostate.[75] At a medical conference in Canada, he exhibits graphic wax models depicting the tissue surrounding the urethra. According to his models, the urethra is nestled in a tangle of glands and prostatic ducts that also empty into it. Though they sometimes surround the urethra entirely, the glands and ducts tend to cluster along its sides and toward the vagina. Huffman provides a catchy description of the urethra and prostate: "The urethra might well be compared to a tree about which and growing outward from its base are numerous stunted branches, the paraurethral ducts and glands."[76]

Huffman reminds us in no uncertain terms that, between individual women, the prostate gland will differ in position, size, and shape. The number of ducts Huffman discovers also varies substantially. In one case, the doctor detects six; in another, he found thirty-one. In some women, the glandular tissue is located near the bladder, while in others, it can be found toward the urethral opening or surrounding the entirety of the urethra. Researchers today cite this morphological variability as the reason for differences in ejaculatory patterns between cis women, why some of these never ejaculate, and why so many medical professionals are still skeptical of the female prostate.[77]

The *Prostata feminina* finds another great advocate in Gräfenberg, who left Berlin and emigrated to the United States by

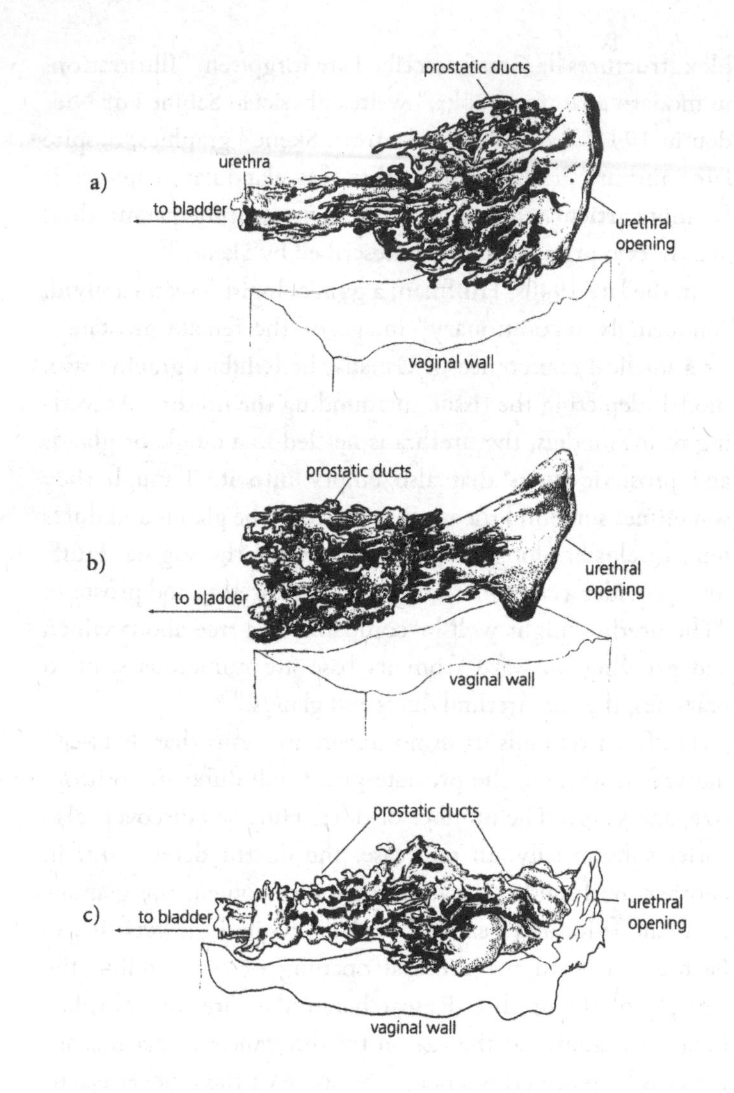

Figure 4.1
The three most common prostatic forms: a) in 66 percent of women, most prostatic tissue is located near the urethral opening; b) in 10 percent of women, the tissue is concentrated near the bladder neck; and c) in 6 percent of women, it covers the length of the urethra.

way of the Soviet Union in the early 1940s. In 1944, the gynecologist identifies an "erogenous zone . . . along the suburethral surface of the anterior vaginal wall."[78] Six years later, Gräfenberg publishes an article in the *International Journal of Sexology* about the role of the urethra in female orgasm. He encourages people to embrace the wide variety of female erogenous zones, emphasizing the front vaginal wall. Stimulating this area—where, and I'm jumping ahead here, the G-spot is located—fires up the female prostate. Gräfenberg's article exudes curiosity, openness, philogyny, and sex positivity. It becomes a milestone in the field of sexology. Three decades later, Beverly Whipple and John D. Perry will name this erogenous zone the "G(räfenberg)-spot" in the doctor's honor.

Like his colleague Huffman, Gräfenberg argues that in all women, the urethra is bounded by erectile tissue that can be stimulated from inside the vagina. The tissue swells in response to sexual arousal, he states, and is pressed downward during orgasm. (A nice explanation for what ejaculating icon Shannon Bell notes, namely that the swollen female prostate will sometimes eject penises and dildos from the vagina.)[79] Gräfenberg establishes a connection between this erectile tissue and female ejaculation. He suggests alternatives to the missionary position—like sex performed a posteriori—that provide a better angle for the penis to stimulate the prostate during heterosexual intercourse. In his closing remarks, he stresses that the area described may be more important than the clitoris and must be taken into consideration when treating orgasmic disorders in women.[80]

More than thirty years pass. Gräfenberg's observations go all but unnoticed. In *Sexual Behavior in the Human Female* (1953), Kinsey writes that women, what with their underdeveloped prostate, do not experience "real" ejaculation. Masters and

Johnson reject the idea of female ejaculation in their vaunted 1966 study *Human Sexual Response*. They speculate that male doctors long misinterpreted female sexual fluids because of their own experiences with ejaculation:

> During the first stage of subjective progression in orgasm . . . a number of women [have described] a sense of bearing down or expelling. Often a feeling of receptive opening was expressed. This last sensation was reported only by parous study subjects, a small number of whom expressed some concept of having an actual fluid emission or of expending in some concrete fashion. Previous male interpretation of these subjective reports may have resulted in the erroneous but widespread concept that female ejaculation is an integral part of female orgasmic expression.[81]

In her seminal work *The Nature and Evolution of Female Sexuality* (1966), US doctor and writer Mary Jane Sherfey (1918–1983) makes no mention of vulval ejaculation. It's an astonishing omission, considering the originality of her research and arguments. Sherfey's contribution is so central to the women's health movement of the 1970s, a key force behind the rediscovery of the female prostate and ejaculation, that I'd like to take a moment to introduce her properly.

MARY JANE SHERFEY

Sherfey's book is revolutionary. Her findings provide new insight into female genitalia, sexual response, and capacity for orgasm. *The Nature and Evolution of Female Sexuality* skewers the notion of an inferior, deficient female gender.[82] Sherfey views the penile and clitoral systems as homologous, meaning "any organ or part of one animal that corresponds in some way to an organ or part of another animal." A licensed psychiatrist, she studies sexual differentiation in human embryos along with the evolution of female sexuality and response. She links findings from evolutionary biology, embryology, gynecology, psychiatry, and ethnology. Among other things, Sherfey explains that "the mammalian male is derived from the female and not the other way around," how male genitals develop from female structures, and why this means phases of arousal (including erection) and orgasm are more or less identical in women and men.[83]

Unlike men, however, women are capable of successive orgasms, peerless in their potential for peaking. Maybe the multiorgasmic nympho is actually the biological norm? In any case, Sherfey makes it clear that women's supposed asexuality or innate frigidity is utter bunk. The human embryo, she notes, is primarily female; at six weeks, the male form splits from the fundamental female form. The "mammalian sexual organs begin existence as anatomically and physiologically female structures." Accordingly, female development is linear, whereas development in males "deviates" from the innately female structure: "Embryologically speaking, it *is* correct to say that the penis is an exaggerated clitoris, the scrotum is derived from the labia majora, the original libido is feminine, etc. . . . For all mammals, modern embryology calls for an Adam-out-of-Eve myth!" Clitoral and penile systems are homologous structures reflected in and back at one another. Each genital feature has a corresponding part in the other sex. Sherfey confirms that the inner elements of the clitoris have never been depicted (suggesting that she is unfamiliar with Kobelt's thorough study of female genital anatomy): "In not one of the works of the comparative anatomists (including the major German reviews) are the deeper-lying clitoral structures described—indeed they are not even mentioned! . . . Because the comparative anatomists and biologists do such a thorough job on every other body system, including the male sexual anatomy, this total omission of the cryptic clitoral structures is of interest."[84]

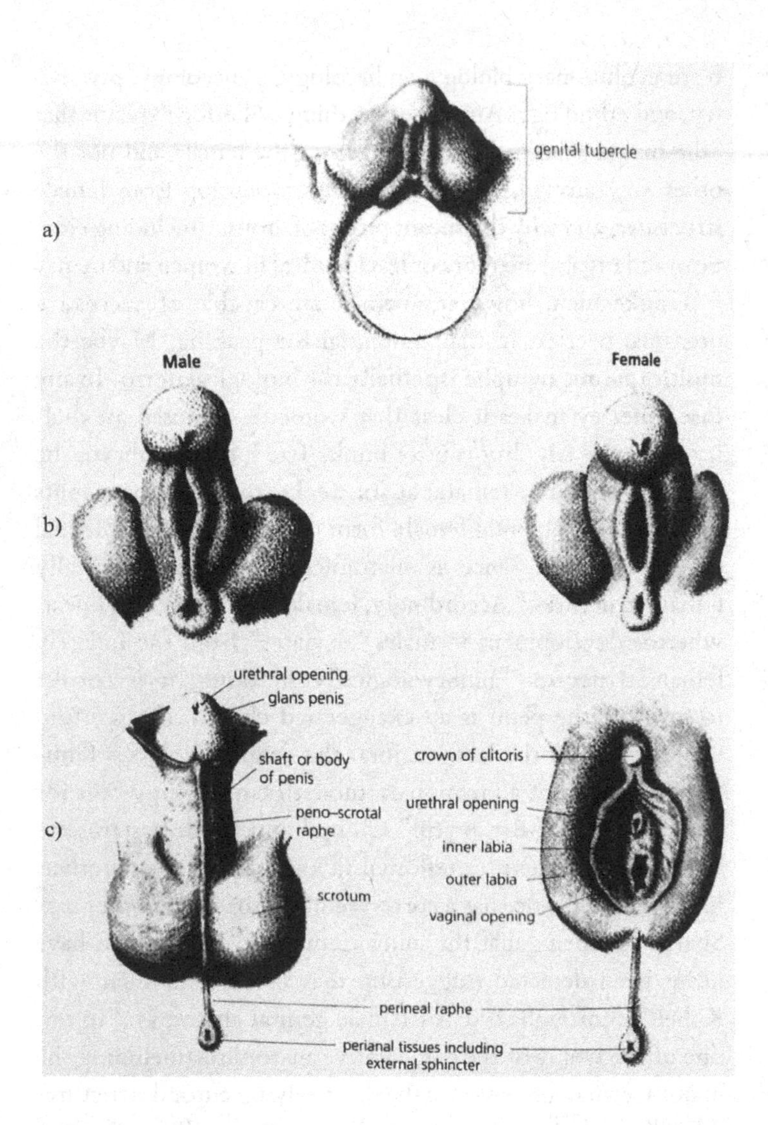

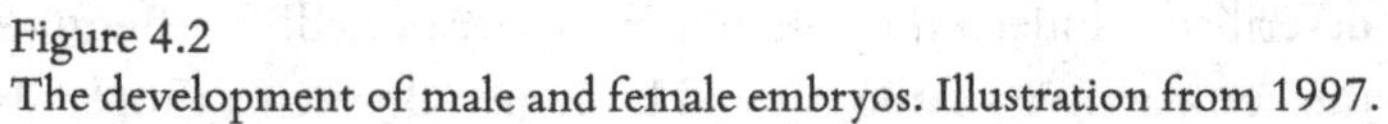

Figure 4.2
The development of male and female embryos. Illustration from 1997.

Sherfey renders a great service in bringing the internal parts of the clitoris to light as well as elucidating the homology of female and male reproductive organs. She thus refutes the assertion that women's genitals are stunted or "castrated" (as per Freudian thought, hugely popular in the 1970s). The clitoris now includes the clitoral body, with the glans or crown at its external tip; a pair of leglike crura; and the *Bulbus vestibuli*, or bulbs of the vestibule. The concealed parts of the clitoris double or triple in size during sexual arousal, making them "only slightly less [significant in size] than the homologous structures in males, relative to body size."[85]

The lower third of the vagina and both inner and outer clitoris function as a unit, which is why "*it is a physical impossibility to separate the clitoral from the vaginal orgasm*." All orgasms are physiologically and qualitatively equal, whether vaginal, clitoral, or otherwise. According to Sherfey, the upper two-thirds of the vagina "[play] no role in either erotogenesis or orgasmic action."[86] She directly compares male and female orgasms, but fails to mention vulval ejaculation. She describes vaginal lubrication ("vaginal transudate") and the fluid produced by the Bartholin's glands. She doesn't have a term for that tissue surrounding the urethra (the female prostate), though. Nevertheless, Sherfey paves the way for the feminist women's health movement and Australian urologist Helen O'Connell, who will be celebrated in the late 1990s for discovering the inner structures of the clitoris.

REDISCOVERIES: FEMALE EJACULATION AND JOY FLOW

Physicians and sexologists have all but forgotten the female prostate and ejaculation until 1978, when Josephine Lowndes

Sevely and J. W. Bennett jog their memory. Sevely and Bennett's essay "Concerning Female Ejaculation and the Female Prostate" presents historical, anatomical, and linguistic evidence in support of their argument, namely that female ejaculation was commonplace until the twentieth century, when it was suppressed because it conflicted with reigning notions of female sexuality.[87] Like Sherfey a decade before, Sevely and Bennett remark on the similarity between male and female sexual response. They state that the female prostate is homologous to the male gland; women, like men, ejaculate; and ejaculation and orgasm do not necessarily occur at the same moment. The authors' recommendation for further research is heeded, and in subsequent years, sexological journals in particular will publish many articles on female ejaculation.

In 1981, *A New View of a Woman's Body*, a "fully illustrated guide" as pioneering today as it was then, offers a new perspective on the subject: ejaculators themselves join the debate.[88] The US-based Federation of Feminist Women's Health Centers publishes the book, which represents years of teaching about sex and self-help. It draws on thousands of conversations, physical examinations, and activists' experiences working at the federation's many day clinics. As such, the book remains one of the most enduring and accessible publications of the lesbian and women's health movement.

New View invites readers to observe themselves and examine their own bodies. It dispels myths about female sexuality and offers self-help techniques. Its detailed illustrations, photographs (including sensational, full-color series on the vulva and cervix), and thorough definition of the clitoris are exceptional to this day. Sherfey's book, "the only modern book that they found to be illuminating," provides the foundation for a new definition of the clitoris.[89] The authors of *New View*

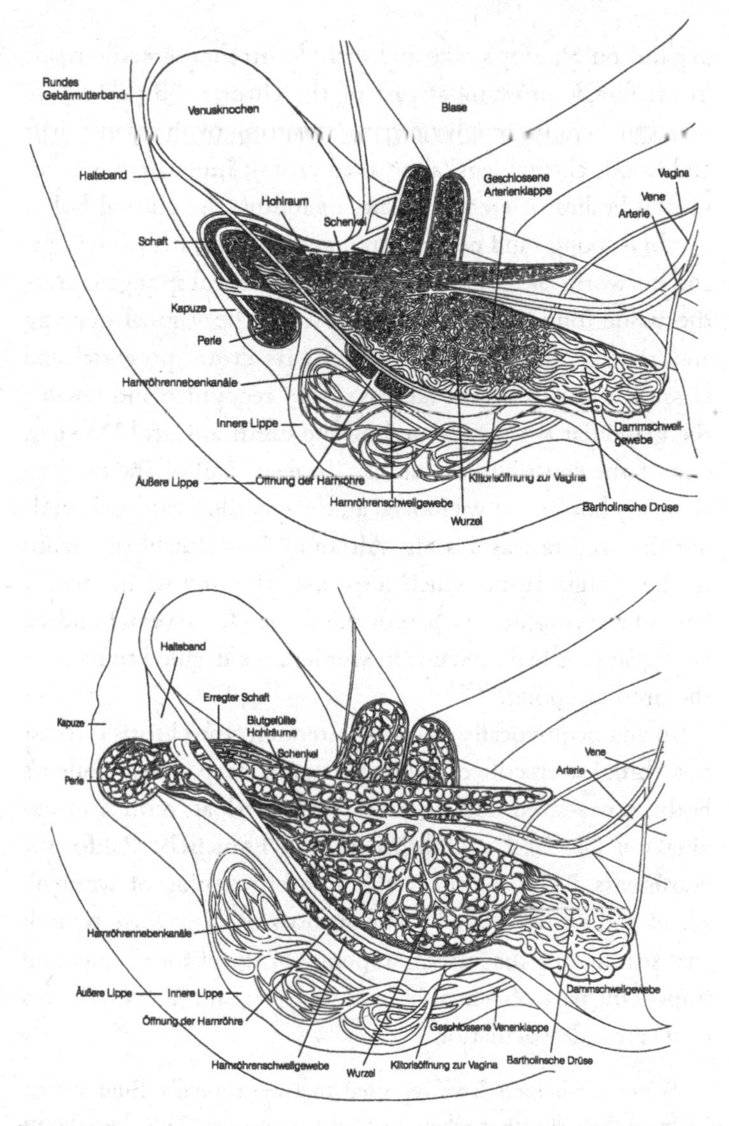

Figure 4.3
Cross sections of the nonerect clitoris (*top*) and clitoris during sexual arousal (*bottom*), from *Frauenkörper—neu gesehen* (1992) (*A New View of a Woman's Body* [1981]).

expand on Sherfey's take and include urethral erectile tissue (read: female prostate) as part of the clitoris: "By *clitoris* we mean the whole complex organ, consisting of the glans; shaft and hood; clitoral legs (also called crura); inner lips; hymen; several bodies of erectile tissue, including the clitoral bulbs, urethral sponge and perineal sponge; muscles; nerve endings; and networks of blood vessels."[90] The perineal sponge is erectile tissue found in the region between the vaginal opening and rectum. The authors highlight its erotic potential and classify it as part of the clitoris. They recognize and rename the glandular tissue surrounding the urethra. Carol Downer, one of the contributors, recalls, "In nearly all of the modern anatomy books that we looked at, the erectile tissue surrounding the urethra was missing. Although it is clearly analogous to the spongy tissue which surrounds the urethra in men, it hasn't been considered a part of the clitoris for several hundred years. Since it had no name in women, we decided to name it the urethral sponge."[91]

In an unequivocally political statement, the clitoris is recast as a complex erectile organ that extends deep into a woman's body, a pea-sized nub no longer. The lesbian erotica magazine *On Our Backs*, published in San Francisco, California, comments, "This is connectable to a broadening of women's social and sexual roles."[92] The authors of *New View* remark that stimulating the "urethral sponge of the clitoris" can be an important part of sexual arousal and orgasm. The book also covers female ejaculation:

> Some self-helpers have reported that occasionally fluid squirts from their clitorises when they have an orgasm. This fluid shoots out like a stream in bursts. One woman described this phenomenon as "gallons" of fluid, distinct from vaginal sweating. Some women have confused this with urination. One woman noted that

> the fluid had an odor different from urine. Unlike the involuntary urination that occurs in a small number of women during sex, this fluid is chemically different from urine and appears to be ejaculated from the paraurethral glands located in the urethral sponge of the clitoris. The same structure . . . becomes the prostate gland in the male, which later contributes to the male ejaculate.[93]

The women's health movement in Germany takes even greater care with language, perhaps, than the Americans, replacing "derogatory, ugly, and male-oriented terms."[94] In the German

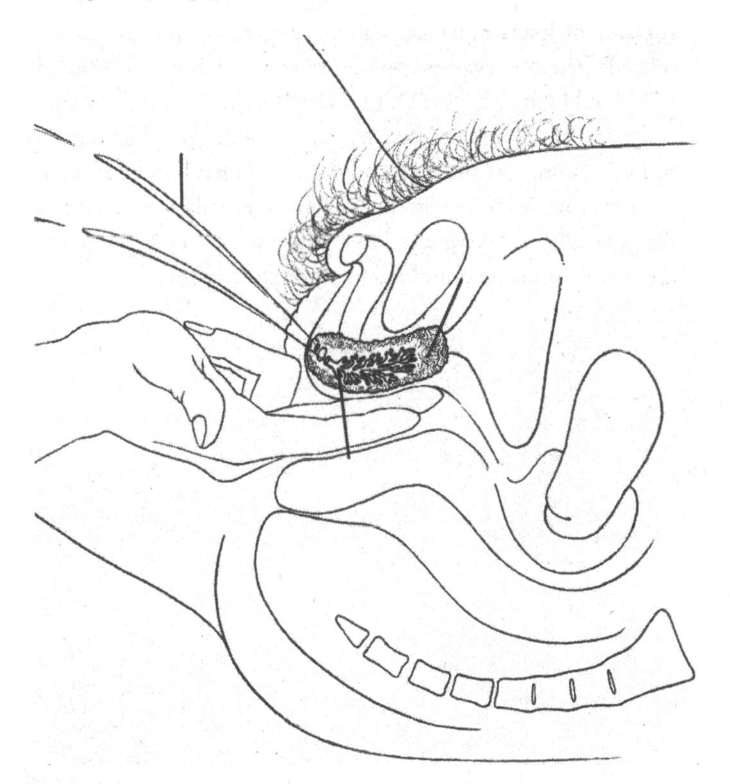

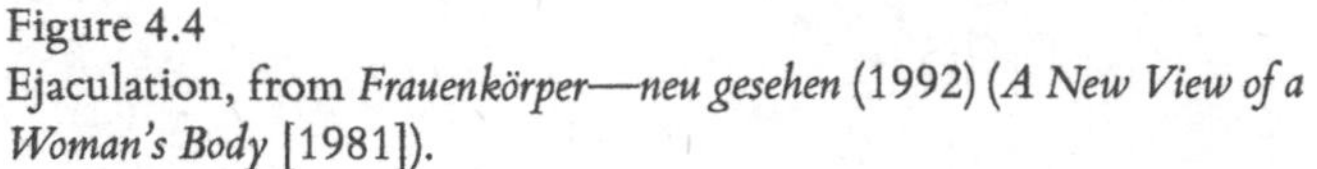

Figure 4.4
Ejaculation, from *Frauenkörper—neu gesehen* (1992) (*A New View of a Woman's Body* [1981]).

rendering, for instance, "female ejaculation" is rechristened *Freudenfluss*, or "joy flow." The new term never catches on, though, perhaps because "joy" doesn't fully reflect orgasmic elation and "flow" fails to capture the characteristic gush of ejaculation.

It's activists from the women's health movement who first alert Whipple and Perry to the existence of vulval ejaculation and urethral erectile tissue, thus inspiring work that would lead to their discovery of the G-spot:

> A group of lesbians, having had this experience [of ejaculation], related their observations to sex researchers Beverly Whipple, R.N., and John Perry, Ph.D. In reviewing the literature, Whipple and Perry found that a researcher named Grafenberg had reported similar findings in the early 1950s. . . . [They] concluded that these women were experiencing female ejaculation and coined the phrase "Grafenberg spot" to identify what they believed to be the site of stimulation and ejaculation in the vagina.[95]

5

ANATOMY REVISITED: G-SPOT, VAGINA, AND CLITORIS

UFO LANDING: WELL STUDIED, POORLY EXPLAINED—THE G-SPOT

> How has it happened that a phenomenon as widely experienced as female ejaculation has not been recognized by the medical profession, and has been dismissed as Victorian pornographic fantasy, or as urinary stress incontinence?
> —Alice Kahn Ladas, Beverly Whipple, and John D. Perry, *The G Spot*

Whipple and Perry team up with Ladas, a psychologist, to explore the lesbian activists' assertions regarding female ejaculation. In 1983, they publish *The G Spot and Other Recent Discoveries about Human Sexuality*. The book is intended for "everyone who is interested in human sexuality," and the authors hope to help "millions of women and men" lead more satisfying sex lives.[1] Indeed, millions do buy the book, which tears through the book club circuit. The authors give interviews and appear on talk shows. Big magazines like *Hustler*, *Playboy*, and the science and science fiction monthly *Omni* report on the book, which becomes an international bestseller translated into more than twenty languages. After *The G Spot* was published, Whipple recalls in a 1987 interview, more than ten thousand women came forward with their stories of ejaculation.[2] The G-spot—behind which the female prostate is found—and female ejaculation had arrived in the mainstream. Though filled with firsthand accounts, *The G Spot* presents few new scientific findings, critics point out. The book prompts unprecedented public and scholarly debate. Never before has an anatomical structure attracted this amount of attention and controversy.

Ladas, Whipple, and Perry want to offer readers a better understanding of their own sexuality. They hope to empower women to discover and delight in their capacity for ejaculation. Self-proclaimed educators, the authors urge readers to masturbate, experiment, and practice to boost the body's potential for sensual pleasure and orgasm. The trio also makes it their business to dismantle repressive notions of female sexuality propagated by the likes of Sigmund Freud, Alfred C. Kinsey, or William H. Masters and Virginia E. Johnson. Ladas, Whipple, and Perry explain that most women are unaware of female ejaculation; many think they're peeing at the moment of climax and try to suppress their orgasm. Shame, stress, and fear are the result. Ignorance of female ejaculation compromises or even destroys many women's romantic lives. The authors endeavor to educate and liberate, but by no means pressure, the women and men reading *The G Spot*: "We want women who ejaculate . . . to know that it is a natural response and that it's okay to enjoy it. . . . We want women who do not ejaculate to feel all right about that, too, to enjoy whatever kind of pleasure they experience and not strive for ejaculation."[3]

Ladas, Whipple, and Perry are unfortunately imprecise in defining the G-spot. The term "spot" is also something of a misnomer. (The term *G-zone* is somewhat more accurate, if one overlooks the irritation that an aspect of female anatomy is named after a male "discoverer.") The illustrations in the book, which depict the zone as a sort of hovering bean, do little to help people locate it. Women and their partners immediately set out in search of the secret spot, often in vain, with "thousands of women probing their vaginal walls, tapping for orgasmic tremors."[4] One US marriage counselor likens it to the search for the Holy Grail.[5] Meanwhile, what the authors define as the G-spot is actually old hat, nothing more than a description of the female prostate, which can be stimulated through the vaginal wall:

> The Gräfenberg Spot lies directly behind the pubic bone within the front wall of the vagina. It is usually located about halfway between the back of the pubic bone and the front of the cervix, along the course of the urethra . . . and near the neck of the bladder, where it connects with the urethra. The size and exact location vary. . . . [It] lies deep within the vaginal wall, and a firm pressure is often needed to contact the G spot in its unstimulated state.[6]

Although acquainted with the works of their predecessors Josephine Lowndes Sevely, J. W. Bennett, and Regnier de Graaf, *The G Spot* authors fail to differentiate clearly between two discrete parts: the prostate itself and the area of the anterior vaginal wall that allows for prostatic stimulation. Since the female prostate can be found along the upper, middle, or lower sections of the urethra, this sensitive vaginal zone also varies. The authors "tend to believe that the area of the G spot includes a vestigial homologue of the male prostate."[7] It's not a button for instant pleasure, they explain, but instead an area that swells when stimulated with firm pressure. And although all women have a G-spot, not all will enjoy how it feels when stimulated.

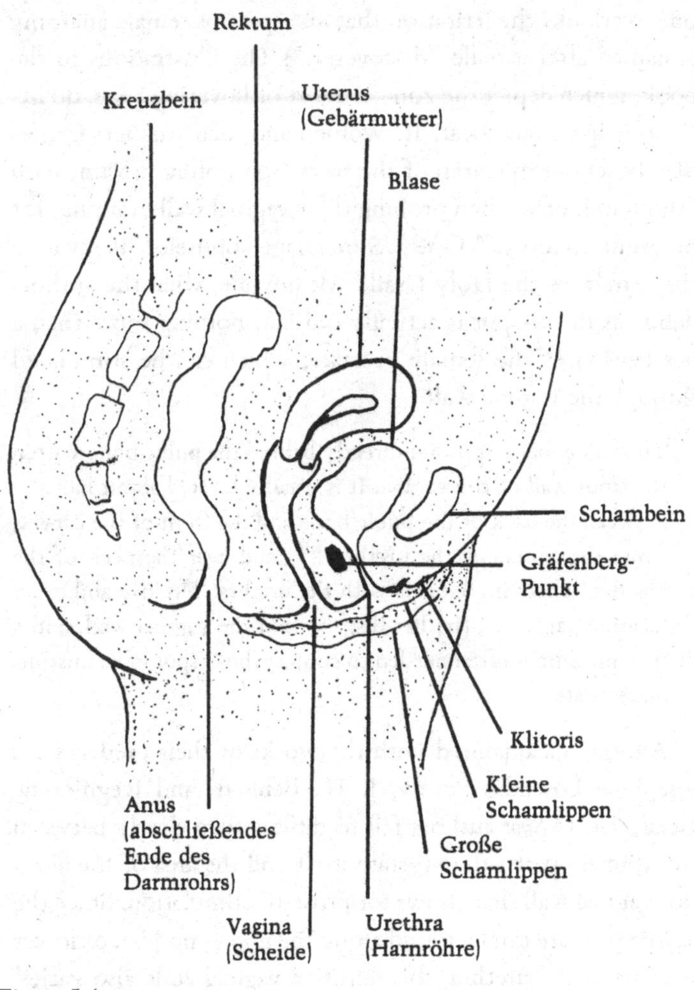

Figure 5.1
The so-called Gräfenberg spot (Gräfenberg-Punkt). Illustration from *Der G-punkt: Das stärkste erotische Zentrum der Frau* (*The G Spot and Other Recent Discoveries about Human Sexuality*) (1983).

When the G-spot is excited, some women will orgasm and some will ejaculate, but ejaculation is also possible without G-spot involvement; Ladas, Whipple, and Perry reckon that between 10 and 40 percent of women ejaculate.[8] The amount of ejaculate varies from person to person and from orgasm to orgasm. Some women report that menses affect their ability to ejaculate as well as the volume and quality of fluid expressed. The authors emphasize the significance of the pelvic floor, arguing that strong pelvic muscles directly influence women's capacity for sexual experience and ejaculation. They urge women to accept "responsibility" for these muscles, to tighten and tone them by doing pelvic floor exercises.

Ladas, Whipple, and Perry decry surgical, medical, or electric interventions. Most surgeries in the mid-twentieth century—"close to fifty varieties of surgical procedures . . . applied to the PC [pubococcygeus] muscle and surrounding areas to control urinary stress incontinence"—are "rather crude." During these procedures, the PC muscle is cut and restitched, in the belief that its truncation and scar tissue will restore muscle tension. Electric therapy for pelvic floor and vaginal muscles is also introduced following the Second World War. Here, electric shocks are administered to activate the muscles. As the effectiveness of electroshock therapy can't be proven either, the authors stick to their recommendation that women practice pelvic floor exercises like those developed by US urologist Arnold Kegel: "It is also cheaper, safer, and more pleasant."[9]

At the 1980 conference of the Society for the Scientific Study of Sex, Whipple and Perry present their ideas on the G-spot and female ejaculation, screening a film that includes an ejaculation scene. Among those in attendance is gynecologist and nonbeliever Martin Weisberg. After seeing the film and chatting with one of the "test subjects," Weisberg is convinced: "I still have no explanation for this, but I can attest to the fact that the Grafenberg spot and female ejaculation exist. Years from now I am sure that a medical school lecturer will joke about how it wasn't until 1980 that the medical community finally accepted the fact that women really do ejaculate."[10]

Ladas, Whipple, and Perry challenge the preferential rank the clitoris holds, questioning the superiority of clitoral climax. They distinguish between three types of orgasm: vulval orgasm triggered by clitoral stimulation, uterine orgasm triggered by vaginal penetration, or, most commonly, a blended orgasm that lies somewhere between the two.[11] The authors thus situate themselves in direct opposition to those titans of sexological research and therapy, Masters and Johnson. Masters digs in his heels at the 1981 annual conference of the American Association of Sex Educators, Counselors, and Therapists, maintaining that "all orgasms involve direct or indirect stimulation of the clitoris."[12]

Ladas, Whipple, and Perry pose fascinating questions about female ejaculation: How does musculature correspond to ejaculation? Do women with strong PC muscles ejaculate more? Are women who ejaculate less susceptible to urinary tract infections (UTIs)? Can women, like men, experience retrograde ejaculation, in which ejaculate flows into the bladder? What's the story with hormones? And how do sexual preference and ejaculation correlate? The authors' preliminary findings suggest that lesbian women may ejaculate more than straight ones, though they call for more research.[13] Cologne-based physician Sabine zur Nieden answers this call in the 1990s and discovers significant differences: among the lesbians she surveys, 42.3 percent have experienced ejaculation, compared to 28.9 percent of the straight women interviewed.[14]

The publication of *The G Spot and Other Recent Discoveries about Human Sexuality* inspires broad discussion, with particular media focus on the elusive spot itself. With everyone busy hunting it down, the chapter on female ejaculation is more or less overlooked. The book's reception comes as no surprise to Deborah Sundahl, whose extensive work on female ejaculation begins in the 1980s; whatever the trio's success, their juggernaut is no match for the taboo against female ejaculation. Another baffling fact is that the women's movement largely ignores the "discovery" of the G-spot and female ejaculation. It turns out that Ladas, Whipple, and Perry's work nettles second-wave feminists in particular, and a quick look back helps us understand why they're loath to join the conversation.

CLITORIS VERSUS VAGINA

Freudian psychoanalytic theory conceives of women as ill-begotten male creatures—as castrated men.[15] In his theories of sexuality and gender, Freud states that boys, girls, men, and women all consider female genitalia inferior to male genitalia, and view the clitoris as a stunted phallus. Little girls, on discovering they don't have a penis, feel "castrated." Over the course of puberty, Freud contends, girls undergo a "transference of the erogenous excitability from the clitoris to the vagina."[16] In other words, the vagina is the primary erogenous zone in women—not the clitoris. Freud devalues the clitoris, elevating the vagina as the main source of female arousal. Healthy grown women derive pleasure from vaginal penetration. This concept "brought the male-centered heterosexual model of sexuality to its phallocentric apogee, and set in stone for the next century that it was not only appropriate but essential for women's sexuality to be defined in terms of male preferences."[17] Freud doesn't deserve the title of "father of the vaginal orgasm," though, as he isn't actually the one to propose two distinct categories of climax.[18] Instead, his adherents advance the notion of an "immature" clitoral and "adult" or "mature" vaginal orgasm.

In the 1950s, researchers like Kinsey, Masters, Johnson, and Shere Hite initiate a gradual paradigm shift: the clitoris is reinstated as an important—if not the *most* important—sexual organ, while the vagina's relevance to female orgasm is called into question. Discourse around sex and sex therapy in the subsequent decades will reject the vagina's capacity to provide pleasure. This rejection is at once a disavowal of the Freudian concept of femininity, which explains the emotionality of certain statements made regarding the vagina in the latter half of the twentieth century.

In 1954, Kinsey describes the vagina as an unfeeling cavity and compares it to the bowels. The "internal (entodermal) origin of the lining of the vagina makes it similar in this respect to the rectum and other parts of the digestive tract." With the exception of the anterior vaginal wall (that is, the side facing the belly and prostate), Kinsey states that the inside of the vagina is "quite insensitive" to touch. There's no male equivalent to the vagina, which "is of minimum importance in contributing to the erotic responses of the female. It may even contribute more to the sexual arousal of the male than it does to the arousal of the female."[19] Masters and Johnson contend that all orgasms are physiologically alike, however triggered. The clitoris, defined only as the crown and shaft, is also stimulated during vaginal penetration:

> This rhythmic movement of the clitoral body in conjunction with intravaginal thrusting and withdrawal of the penis develops

> significant secondary sexual tension levels. It should be emphasized that this same type of secondary clitoral stimulation occurs in every coital position, when there is full penetration of the vaginal barrel by the erect penis.[20]

Hite's international bestseller, *The Hite Report*, published in 1976, is based on a survey of 3,019 women between the ages of fourteen and seventy-eight. Hite finds that barely 30 percent of those surveyed achieve climax by vaginal penetration alone; one in every five (19 percent) requires clitoral stimulation on top of penetration. The survey also shows that women can orgasm just as quickly as men: "It is, obviously, only during inadequate or secondary, insufficient stimulation like intercourse that we take 'longer.'"[21]

The female body and sexuality are central concerns of second-wave feminism. When it comes to their own bodies and desires, women are now in charge. The "nether regions," long thought dangerous or dirty, are now being explored, exposed, described, and, if necessary, renamed. Women rediscover forgotten or suppressed knowledge, spearhead research, and cast off feelings of shame or guilt. There's a reason one of the most successful books of the time is called *The Shame Is Over* (1976), by Dutch writer Anja Meulenbelt. Women are demanding more space in bed and more fun. Private life has become political, as sex always has been. "Sexuality is at once the mirror and instrument of women's oppression in all areas of life," Alice Schwarzer, a German journalist and vocal feminist, writes in 1975.[22]

The movement condemns Freudian concepts of femininity, the pressure to engage in "normal" penetrative intercourse, heteronormativity, and the way supposedly "frigid" women are pathologized. In *The Myth of the Vaginal Orgasm*

(1970), Anne Koedt, a US radical feminist, writes that vaginal functions are limited chiefly to "1) menstruation, 2) receive penis, 3) hold semen, and 4) birth passage."[23] The vagina is ultimately a sexual organ "for" men, who use it to achieve orgasm, Koedt contends, thus alluding to Simone de Beauvoir, who as early as 1949 associates vaginal penetration with assault in *The Second Sex*: "The woman is penetrated and impregnated through the vagina; it becomes an erotic center uniquely through the intervention of the male, and this always constitutes a kind of rape."[24]

The clitoral crown dotted with thousands of nerve endings becomes a metaphor for female autonomy. It becomes women's most important sexual organ. The clit has just one job: to provide pleasure. Schwarzer says vaginal orgasm doesn't and can't exist because the vagina has about "as many nerves as the large intestine, which is to say, almost none." The less favorably women view vaginal penetration, the more weight men attach to it. In *Der "kleine Unterschied" und seine großen Folgen* (The "minor difference" and its major consequences), published in 1975, Schwarzer goes so far as to declare vaginal penetration the basis of female oppression and male dominance:

> Our sexual norms as dictated are clearly about female subjugation and male exercise of power. . . . The myth of vaginal orgasm (and with it, the significance of penetration) is what secures men's sexual monopoly over women. And the sexual monopoly is what secures men's private monopoly, which itself provides the foundation for male-dominated society's public monopoly over women. . . . This is why, in order to upset gender roles, the male sexual monopoly must be destabilized from the ground up.[25]

Vaginal intercourse thus symbolizes patriarchal appropriation, power, and dependence. It represents an active, subjugating male force and passive, subjugated female counterpart. Women zealously (re)appropriate the clitoris and "clitoral orgasm." In 1971, Carla Lonzi, a cofounder of Italy's first feminist collective, Rivolta Femminile (Female Revolt), associates clitoral and vaginal arousal with contradictory types in her internationally acclaimed essay "La donna clitoridea e la donna vaginale" ("The clitoridian woman and the vaginal woman"): "The 'vaginal' woman is one who, in captivity, has been thought to give pleasure to the patriarch. The 'clitoral' woman is one who disobeys the emotional pressures of integrating with the former (the types of pressures that have an effect on the passive woman) and expresses herself with a sexuality that does not coincide with coitus."[26] The vagina is in cahoots with men; it's the fifth column of the patriarchy within the female body. It is a numb, conformist, adaptable tube. Though the reasons for vaginal antipathy are understandable, it's depressing to remember the abhorrence directed toward this part of the female body.

One outlier is Australian author, publicist, and feminist Germaine Greer. She vehemently objects to how the feminist movement treats the vagina. In *The Female Eunuch* (1970), she writes, "Unhappily we have accepted, along with the reinstatement of the clitoris after its proscription by the Freudians, a notion of the utter passivity and even irrelevance of the vagina."[27]

THE WOMEN'S MOVEMENT KEEPS MUM

Context helps explain why the women's movement largely ignores Ladas, Whipple, and Perry's findings and recommendations for renewed study of the vagina. Women fear *The G Spot* is establishing new sexual standards (female ejaculation!) and resurrecting old ideals (vaginal penetration, vaginal orgasm!). The authors' indirect depiction of vaginal orgasm as more intense or "deeper" than "superficial" clitoral climax ruffles feathers.[28] Among the loudest critics is Barbara Ehrenreich, a US author

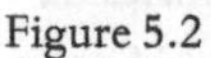

Figure 5.2
"You can't pee against a wall, though, and you never will!" Caricature from the 1970s.

and political activist, who decries the authors' attention to vaginal orgasm and Kegel exercises; of all body parts to tone, women's PC muscles are those that "hold the penis in place."[29]

There are plenty of women who don't like the idea of female ejaculation either. Is it really necessary to come "like a man"? Aren't women in the middle of discovering their own distinct sexuality that diverges from male standards and sway? Surely female ejaculation boils down to misogynistic male fantasy.[30] *EMMA*, founded in 1977 by Alice Schwarzer as a magazine "for women, by women," exemplifies feminists' reticence toward female ejaculation, the prostate or urethral sponge, and the G-spot. The bimonthly ignores these topics through the 1980s and 1990s. One exception is an article from 1987 by zur Nieden, whose research leads to her 1994 publication *Weibliche Ejakulation. Variationen zu einem uralten Streit der Geschlechter* (Female ejaculation: Variations on an ancient battle of the sexes).[31] Berlin's lefty feminist journal *COURAGE* doesn't touch it either. In 1995, *clio*, a women's health journal published by the Feministisches Frauen Gesundheits Zentrum—an independent institution in Berlin that has offered holistic women's health care since 1974—features an interview with sex researcher and writer Bea Trampenau. When asked why the women's movement has ignored female ejaculation, Trampenau recalls:

> It's in part because of the moral discussion surrounding vaginal orgasm. We women had to discover the clitoris, and we wanted to. It was also very important that we draw attention outward: we can experience a tremendous amount of pleasure without cock, without penetration. The gush comes from us. It's reminiscent of penetration and the associated arousal, but we've moved beyond that and can stand by it now. We can say: there's a lot inside of us we don't ever want to forget.[32]

"TIMES HAVE CHANGED": SEXOLOGY OF THE 1980S DISCOVERS EJACULATION

The shock waves unleashed by *The G Spot* rattle the field of sexology in the United States, which for decades honored the "paradigm of cliterocentricity," largely overlooking the vagina.[33] US psychologist Cynthia Jayne confirms this blind spot in her 1984 essay, "Freud, Grafenberg, and the Neglected Vagina: Thoughts Concerning an Historical Omission in Sexology." With critical self-reflection, Jayne asserts that female ejaculation and the G-spot deserve further study:

> The vagina was not expected to demonstrate such sensitivity, and thus no serious programs of research were pursued on vaginal responsiveness. Until recently, sexologists perhaps "foreclosed" on the data too soon. The concepts of female ejaculation and Grafenberg spot stimulation deserve further attention in sex research, as does the general issue of vaginal responsiveness and sensitivity. . . . Finally, as scientists, we must be aware that the truth as we comprehend it can, at times, be a restrictive self-fulfilling prophecy.[34]

Over the course of the 1980s, many studies take up Ladas, Whipple, and Perry's work, proving or disproving, classifying, and expanding on their arguments. The questions examined during that decade include: Where does the fluid come from? What's it made of? How do ejaculate and urine differ? Does female ejaculation correspond to male ejaculation? What role do pelvic floor muscles play? Will the rediscovery of ejaculation enrich people's sex lives or introduce new pressures for sexual performance? Female ejaculation is debated most actively in the United States. Here's an overview of the most important voices and theories at the time.

In 1984, New York–based psychiatrist and psychoanalyst Desmond Heath returns to an article on female ejaculation he was unsuccessful in getting published back in 1979. He's amazed by the shift in attitudes in the intervening five years:

> The climate of science has changed since 1979 when female ejaculation was deemed at best a projection of male Victorian pornographers . . . , and at worst an idea that would cast suspicion on the scientific integrity of the holder. Times have changed. A new climate of opinion appears to have allowed a dormant idea to arise spontaneously, as if new, in more than one place at a time.

Heath shares the case of an analysand who reports ejaculating a strange fluid. He digs for material that might explain the woman's emissions and comes across the female prostate and ejaculation. He finds the works of Alexander Skene, John W. Huffman, and Ernst Gräfenberg, calls the Masters and Johnson Institute ("five minutes on the phone with Masters convinced me that the knowledge had never been lost for it had never been known"), confers with an anatomist at Mount Sinai School of Medicine, and ultimately conducts his own study.[35] He examines eighteen urethras, all of which feature glandular tissue in various forms. Heath detects prostate-specific antigen (PSA) in fifteen of the samples and prostatic acid phosphatase in twelve. This immunohistochemical study establishes the homology of male and female prostates, though Heath maintains it's not yet sufficient proof of female ejaculation.

Studies in the 1980s focus on distinguishing between ejaculate and pee, with researchers looking for elements present in male prostatic secretions but not in urine, like PSA. Nearly all conclude that female ejaculate is not urine. There is also discussion as to whether the term "ejaculation" is appropriate for both men and women.[36] Joseph G. Bohlen expresses misgivings. In his 1982 essay "'Female Ejaculation' and Urinary Stress Incontinence," the US physician critiques the use of "ejaculation" for most any fluid discharged during intercourse and questions the paraurethral glands' capacity for producing substantial amounts of fluid.[37] To his mind, sexologists are working hard to establish a new myth, namely that of female ejaculation.

In 1984, Edwin G. Belzer, Whipple, and William Moger publish a brief article in the *Journal of Sex Research*, titled "On Female Ejaculation."[38] After summing up recent studies, the authors present their own, in which they determine ejaculate and urine to be distinct substances. In the same issue, Perry and Whipple publish a study of the connections between PC muscles, ejaculation, and orgasm; they write that, compared to women who do not ejaculate, those who do tend to have stronger pelvic floor muscles and more pronounced uterine contractions.[39]

Heli Alzate, a professor of sexology from Colombia, and Zwi Hoch, an obstetrics and gynecology specialist at Tufts University, recapitulate the heated debate in their 1986 article, "The 'G Spot' and 'Female Ejaculation': A Current Appraisal." They ask readers to bear in mind that female ejaculation, itself a semantically confusing term, doesn't serve any function. They also object to the use of "female prostate" for the Skene ducts. Although the G-spot cannot be verified, Alzate and Hoch are willing to accept that women "possess a zone of tactile erotic sensitivity on the anterior vaginal wall, which in many of them extend to the entire anterior wall and to the posterior vaginal wall. It also seems that some women emit a fluid through the urethra at orgasm, although its true nature and anatomical origin are still unclear."[40]

In 1987, Heath publishes another article, this time in the journal *Medical Hypotheses*.[41] Even in young girls, he writes, the prostate is a functioning organ that can effect ejaculation in response to early childhood masturbation. In a 1989 editorial, Marc A. Winton notes that the concept of female ejaculation offers relief to women who worry they're peeing during sex. On the other hand, it also puts pressure on women who don't ejaculate. Winton remarks that gender roles might change in response to common acceptance of ejaculation, the female prostate, and women's capacity for multiple orgasms:

> Furthermore, it appears that the G-Spot and female ejaculation may actually confound issues of control. While the male has traditionally been defined as having a stronger sex drive and being the leader in sexual relations, the G-Spot, female ejaculation, and multiple orgasms not only reaffirm the notion of women as sexually superior to males, but also might put additional pressures on men to sexually perform in order to satisfy women.[42]

Karl F. Stifter's book *Die dritte Dimension der Lust. Das Geheimnis der weiblichen Ejakulation* (The third dimension of pleasure: The mystery of female ejaculation) comes out in Germany in 1988. In it, the Austrian psychologist and sexologist evaluates a wide range of materials, providing new translations for greater accessibility. He conducts his own studies of urine and ejaculate samples.[43] The substances are demonstrably different, he concludes; ejaculate is "indisputably" a glandular secretion. Stifter concedes that other "ejaculatory sources" must exist beyond the prostate: "It is hard to imagine the female prostate could produce such large amounts of fluid on its own." He also recommends further research into the "mechanics of the ejaculatory trigger."[44]

In the late 1980s, 2,350 US and Canadian women participate in a survey on female ejaculation and the G-spot. Reportedly 39.5 percent have ejaculated during climax, while 60.5

percent have not. What's interesting is that nearly 60 percent of the respondents have heard of female ejaculation, whether from books, medical journals, magazines, or movies. Women are talking too, with 56.1 percent of those surveyed admitting to having discussed the topic with friends or family. The numbers show that by the end of the decade, female ejaculation is known to the broader public. Among the women who have experienced the phenomenon themselves, however, 10.3 percent have sought psychiatric care for it. The authors of the study explain that ejaculation is a source of stress to many women, who worry they're actually urinating.[45]

Francisco Cabello Santamaría makes another intriguing contribution to the debate in 1997. Curious as to why all women have a prostate while only some ejaculate, the Spanish psychologist and sexologist tests the urine of twenty-four women (six of whom ejaculate) before and after orgasm. He's looking for PSA. He also tests those six women's ejaculate for PSA. Cabello Santamaría detects PSA in 75 percent of the postcoital urine samples and all six ejaculate samples. His conclusion: during orgasm nearly all women produce a fluid originating in the female prostate, at least in part, hence the presence of PSA. In some women, however, the fluid isn't expelled through the urethra. Instead, retrograde ejaculation occurs, as it does in men, in which the fluid returns to the bladder, where it mixes with urine and is peed out. Women may also emit so little ejaculate, they don't even notice. With his study, Cabello Santamaría hopes to reassure ejaculating and nonejaculating women alike, and "break the growing myth of the 'ejaculating superfemale'": first of all, it's normal to ejaculate, and second, at about 75 percent, nearly all women do.[46] Those who don't come visibly (and may be desperate to) are either ejaculating into the bladder or releasing an imperceptible amount of fluid.

As ever, experts disagree about the amount and composition of female ejaculate. In 2011, biologist Alberto Rubio-Casillas and medical doctor Emmanuele A. Jannini conclude that women emit two fluids from the urethra during orgasm: a thin, clear, odorless fluid (approximately 57 to 120 milliliters) that gushes or squirts out, and ejaculate (about 0.50 to 0.90 milliliters), a viscous, milky liquid with high PSA values. The two might form a single stream. The presence of two fluids may explain the wide-ranging research outcomes. "All previously published studies [fail] to differentiate these two fluids, thus generating confusion and controversial findings," the researchers declare.[47] In a 2017 narrative review, Czech physicians Zlatko Pastor and Roman Chmel also determine that women ejaculate two fluids: ejaculate from the prostate containing PSA, and squirting liquid from the bladder containing uric acid, urea, and creatinine. Female ejaculation and squirting are different mechanisms that release different fluids produced in different organs.[48] Pastor and Chmel recommend further research.

In 2021, an international research team publishes its analysis of forty-four accounts of female ejaculation from 1889 to 2019. "It became apparent," the authors state, "that clinical and anatomical studies conducted during recent decades provide substantial evidence in support of the female ejaculatory phenomenon." Confusing, contradictory findings in the study of female ejaculation, they continue, may come down to the fact that these studies are actually considering different substances: female ejaculate, squirting fluid, or even vaginal lubrication. Consequently, "proper fluid identification and use of terminology will ensure that future research in this area is not futile."[49] Meanwhile, male squirting is a thing now too; the first scientific study on that topic comes out in 2018.[50]

MALE CLITORIS AND FEMALE GLANS: A RADICAL NEW VIEW OF THE ANATOMY OF GENDER

A decade after her essay "Concerning Female Ejaculation and the Female Prostate" helped revive debate on the subject, in 1987 Sevely publishes a book on her many years' research. Taking an interdisciplinary approach, she proposes a new take on anatomy and biology, conducts her own study on female ejaculation, and develops a radical theory of the extensive homology of male and female genitalia. Sevely's especially interested in the one sexual phenomenon that supposedly differs in men and women: ejaculation.

She finds it unlikely that women don't ejaculate, not least because of countless historical references to female sexual fluids. In *Eve's Secrets: A New Theory of Female Sexuality*, Sevely advances a new perspective on anatomy, and in particular, calls for a fundamental redefinition of the clitoris and vagina. She demonstrates that men have a clitoris inside the shaft (*Corpus spongosium*, the erectile feature that fills with blood, causing the penis to stiffen and stand), and women have a prostate and female glans in addition to the clitoral crown. According to Sevely, the vagina corresponds to the penis, "female erection is a reality," and both men and women ejaculate.[51] She calls the tip of the male and female clitoris the "Lowndes crown." This is a first—a part of human anatomy named after a woman!

Sevely rejects the dualistic concept of clitoris versus vagina. She explains that, sexually speaking, the clitoris, vagina, and urethra constitute a unified whole. There's only one type of orgasm, though it may be triggered by different areas. She dubs the preorgasmic interplay of clitoris, urethra, and vagina the "C. U. V. response." The vagina consists of the Lowndes crown, clitoral corpus and two crura, spongiosum (female prostate), vaginal musculature (the perineal muscles, urogenital diaphragm, and pelvic diaphragm), and vestibular bulbs.[52] Sevely defines the tissue around the urethral opening as the "woman's glans."[53] She also recognizes the urethra, surrounded by the female prostate, as a sexual organ. The floor of the urethra is anatomically inseparable from the vagina. Together they form a wall: the urethral floor is at once the vaginal ceiling. The prostatic glands are embedded along this part of the vagina.[54]

In adult women, Sevely writes, the urethra and prostatic glands swell when sexually stimulated, which is something one can easily feel "just inside the top of the vaginal opening." The surface of this touchable region "is composed of tiny ridges and furrows called rugae (Latin for 'folds'), which are most pronounced in the groovelike spaces to either side of the bulge and may extend to the floor of the lower vagina."[55] Isn't she just describing the G-spot, albeit far better than Ladas, Whipple, and Perry managed?

Given that male and female genitals correspond down to the last part, Sevely contends, it goes without saying that male and female sexual response will as well. Climax follows a standard pattern: contractions, orgasm, ejaculation, and deflation.[56] Sevely accepts the theory that "all women normally ejaculate sexual fluids from the prostatic glands." In one case study, she and her team find that female test subjects produce up to 126 millilеters (about 4 ounces) of fluid, whereas men typically ejaculate 2 to 6 millileters of semen. The urea content of female prostatic fluid is "at a much lower range" than what's found in urine, and the ejaculate has its own characteristic odor. In addition to the fluid women ejaculate in "a series of 'spurts' or 'jets'" during coitus, Sevely echoes an observation made by de Graaf three centuries earlier: she notes that prostatic fluid sometimes "gushes" from the urethra during the initial stages of arousal. In her conclusion, Sevely optimistically states, "With barriers removed and acceptance of the fact that female ejaculation is normal, women, as many have already reported, find their psychological well-being enhanced by a sense of confidence rather than doubt about a physiological phenomenon which until very recently was usually diagnosed as incontinence, or automatically associated with injury or infection."[57]

Like Gräfenberg before her, Sevely deems the urethra a sexual organ. Several years later, zur Nieden will demonstrate that the medical desexualization of the female urethra is one reason ejaculation was forgotten or misdiagnosed as pathological incontinence. Research was further hindered by the fact that two siloed medical specializations—urology and gynecology—were responsible for the same group of organs. These days, the area is known as the urogenital system, a semantic umbrella that takes into account the genital and urological organs' shared embryological origin. The name also denotes the direct link between the clitoris, vagina, uterus, Fallopian tubes, erectile features, prostate, urethra, and bladder, and the way disease or disorder in one will typically affect the others.

Back to Sevely and her impressive stance on the homology of male and female genitalia. She explains that prostatic function is influenced by a number of factors: the size and number of ducts, sensitivity of the tissues, age, the manner and frequency of stimulation, and the woman's arousal. Sevely also reminds us that the vagina, "far from being a passive space, is a complex entity of active space and deeper-lying sexual parts." She outlines vaginal activities that heighten orgasmic response and clarifies the distinct changes that occur to the vaginal canal as various parts of it swell. The "'clasping,' 'caressing,' 'kissing' actions of the middle and lower vagina" make it so much more than an inert tube formed by the penetrating member.[58]

To describe these vaginal activities, about thirty years later German writer and artist Bini Adamczak coins the term "circlusion" as a counterpart to penetration: "Both words signify the same material process, but from opposite perspectives. Penetration means to insert or stick in, whereas circlusion means to enclose or pull on. That's it. What this does, though, is reverse the relationship between activity and passivity."[59] The neologism brings to light something women and their partners have long experienced yet struggled to name: the way the vagina will suck, grab, and hold onto things.

Sevely's presentation of male and female genitalia befits the period in which it was published, when the demand for gender equality was growing in political and private spheres. The author hopes her concept of anatomical and functional symmetry between the sexes will "help men and women to understand both themselves and each other better," and lead to relationships characterized by equality.[60] For centuries, anatomy has provided "proof" of women's inferiority, but now Sevely has developed a vision of equality reflected in genital symmetry. Homology of the clitoral complex, penis, and sexual responses anchors gender equality inside these human bodies. Ejaculation is at once proof and symbolic of this parity.

MYTH, CASH COW, AND P-SPOT: THE G-SPOT, THEN AND NOW

For a brief spell in the late 1980s, the G-spot was a concept familiar to laypeople and medical professionals alike. Barely twenty years later, though, the skeptics are back at it. Physicians and sexologists start calling the G-spot a myth, a "gynecologic UFO."[61] Some suggest "P-spot" (P, as in placebo) as an apter alternative.[62] Others, meanwhile, embrace the G-spot as a lived, researched, scientifically proven fact. In 2009, the *Journal of Sex Medicine* hosts a conference on the G-spot. A wide array of opinions are presented and later published in a paper, "Who's Afraid of the G-Spot?" The conclusions are pretty sobering: even after decades of research, further research is required. One of the "most challenging aspects of female sexuality" remains a mystery.

Although not yet confirmed as an anatomical feature, this erogenous zone sparks a multimillion-dollar industry.[63] Books and videos explain what the G-spot is and how to find it. Special dildos and vibrators promise G-spot-specific stimulation. There's also a boom in genital plastic surgery. In 2002, David Matlock, one of the best-known cosmetic gynecologists in the United States, introduces the G-Shot®. The procedure, also known as G-Spot Amplification® and available in the United States and Europe, is a form of G-spot augmentation. Hyaluronan, which acts as a filler, is injected into the anterior vaginal wall. This lunchtime procedure, the Los Angeles–based surgeon's website states, "can be life-changing for women and their sex-lives."[64] The G-Shot® is for everybody, Matlock assures potential clients; even sexually satisfied ladies can benefit from a little boost. Doctors and other experts have called the uptick in genital plastic surgery an "alarming trend."[65] Despite such warnings, a price tag of anywhere between $1,000 and $2,500, and waning effects after just three to five months, G-spot augmentation is in demand. In the history of genital plastic surgery—a field perhaps best characterized by its absurdity—this particular service represents a new low.

Whipple herself speaks out against G-spot augmentation. She contributes the entry on the G-spot for the 2015 *International Encyclopedia of Human Sexuality*: "The Gräfenberg spot, or G spot, is a sensitive area felt through the anterior wall of the vagina. It is usually located about halfway between the back of the pubic bone and the cervix, along the course of the urethra. It swells when it is stimulated, although it is difficult to palpate when in an unstimulated state."[66] In other words, the G-spot is the female prostate, or the glandular tissue surrounding the urethra, which can be stimulated through the front wall of the vagina. Thirty years after Whipple and her colleagues Ladas and Perry published their bestseller *The G Spot*, we've wound up at the female prostate again.

BRIEF ASIDE: THE FEMALE PROSTATE, PART II

The Federative International Committee for Anatomical Terminology (FICAT), founded in 1989, brings a group of experts together to assess human anatomical nomenclature. Formed by the International Federation of Associations of Anatomists, the committee's objective is to establish standardized medical terminology for the whole world.[67] In 2001, FICAT updates *Histological Terminology* with the term "female prostate," and instructs medical professionals to stop using "gland," "paraurethral gland," or "Skene's gland" in connection with the organ. It's a real coup.[68]

Slovakian gynecologist Milan Zaviačič (1940–2010) is a driving force behind the change, having studied the organ's structure and functionality since 1980 as well as published more than 500 scientific articles on the subject.[69] Between 1985 and 1999, Zaviačič and his colleagues at Comenius University in Bratislava perform more than 150 autopsies and examine more than 200 patients for a better understanding of the anatomy, histology, and pathology of the urogenital system.[70] In 2006, Radio Prague International applauds Zaviačič for having "discovered" the female prostate, which he proves isn't a vestige of the male gland but instead a functional organ in its own right.[71]

Zaviačič also insists on the correct terminology: female prostate. Calling it the Skene's gland, he argues, suggests a totally different structure and obscures its homology with the male prostate.[72] The female prostate is not found *around* the urethral wall but rather *in* it. On average in grown women, the organ weighs just shy of 0.2 ounces and measures 1.3 by 0.8 by 0.4 inches.[73] Its shape varies significantly between individuals, which Zaviačič explains is what makes identification so hard. He goes on to define six prostatic types. Most common is the anterior or meatal type: in nearly 70 percent of women, the prostate is located in the distal half of the urethra, the part farther away from the body's center. This prostatic type is an important erogenous zone that contributes to female orgasm.[74]

Female and male prostates have the same histological structure, both comprised of glands, ducts, and smooth muscle cells. Like its male homologue, the female prostate has at least two jobs: the exocrine production of female prostatic fluid ("exocrine" indicating a secretion released outside its source) and certain neuroendocrine functions, though research into the latter is in its infancy. (One known process is the production of serotonin by neuroendocrine cells found there.)[75] Zaviačič blames the "chronic lack of interest in this organ among urologists, gynecologists, gynecological urologists, and pathologists" on the fact that a pathology of the female prostate is so long in coming.[76] Cancer, prostatitis (female urethral syndrome), and the benign enlargement of the prostate can afflict women too. Disease in the female prostate is far less common or severe than in men, which may explain Big Pharma's reluctance to bankroll studies of the organ.

As for its role in sex, female prostatic secretions constitute the lion's share of ejaculate, which is evacuated through the urethra. During orgasm, rhythmic local muscle contractions squeeze the contents of the prostate through ducts leading to the urethra. The time has come to shelve the persistent controversy surrounding the interplay of female prostate and ejaculation, Zaviačič writes in 2002: "Having proven the presence of prostatic components (PSA, in particular) in female ejaculatory fluid, we have determined that the female prostate is a primary source of urethral fluid emissions." Ejaculation is brought about, "normally without trouble," by stimulating the G-spot. That said, not even Zaviačič can precisely define that spot in morphological or anatomical terms.[77] He does, however, investigate his belief that the prostate sheds fluid continuously, even when women aren't aroused. He theorizes that these secretions might play a role in reproduction. The fructose found in female ejaculate, for instance, might affect spermatozoal mobility.

"Who'd have thought there could be so many open questions regarding macroscopic human anatomy? And yet, the female prostate polarizes experts to this day," Florian Wimpissinger comments in 2007. The Vienna-based urologist has been researching the female prostate and ejaculation for a long time. "It's interesting that specialists versed in anatomy and surgery—experts in the fields of urology and gynecology as well as anatomy—are usually unable to confirm the existence of the female prostate with any degree of certainty."[78] Many remain skeptical, Wimpissinger argues, because prostatic tissue varies so significantly between individuals. It's difficult to reconcile this variance with the fact that the male prostate always looks the same and is found in the same spot.

Wimpissinger cites a case study of two women, ages forty-four and forty-five, who ejaculate during orgasm. Ultrasounds, biochemical analyses, and urological endoscopies indicate the presence of glandular tissue around both women's urethras. He also detects a clear difference between ejaculate and urine, ultimately determining that female ejaculate is biochemically identical to male prostatic secretions.[79] In another study using MRIs, Wimpissinger identifies the prostate in six out of seven female test subjects.[80] Further parallels, even at the embryological level, exist between the male and female prostate, he explains. Both emerge from the urogenital sinus. In male embryos, the prostate develops under the influence of the androgen dihydrotestosterone, whereas the hormonal profile in female embryos prevents such development. Wimpissinger reaches the same conclusion as countless researchers before him: women's glandular tissue corresponds to the male prostate, and female ejaculation "seems to be more common than generally recognized."[81]

The urologist operates under the assumption that there's a connection between prostatic form and ejaculation. He concludes, "Based on historical accounts, scientific data, and our own findings, discussions on the existence of the female prostate can be reduced to the nomenclature of paraurethral anatomy (glands) and the variability of this gland's embryological development."[82] Like his predecessors, Wimpissinger calls for comprehensive research. Are UTIs and lower urinary tract symptoms caused by a prostatic condition? Is there a correlation between orgasmic intensity and ejaculation? Does ejaculation cause extra-intense orgasms, or is it the other way around?

Another group of Austrian researchers, led by Vienna-based gynecologist Wolf Dietrich, publishes a study in 2011 that discovers functional glandular tissue around the urethra in 50 percent of women. The tissue is most often located along the lower sides of the urethra, near the urethral opening.[83]

The Federative International Programme for Anatomical Terminology (FIPAT), FICAT's successor, revisits the female prostate in the 2019 edition of *Terminologia Anatomica*. The nomenclature experts decide it's best to restore past designations for the organ, now known officially as the "para-urethral glands of female urethra," with "Skene's glands; Prostata feminina; Female prostate" provided as alternatives. FIPAT includes the "para-urethral ducts of female urethra" (synonym: "Skene's ducts") in the retrograde renaming as well.[84] The female prostate remains a contested thing.

FROM CLIT TO COMPLEX: THE CLITORIS MAKES ITS GRAND ENTRANCE

Helen Elizabeth O'Connell causes quite the stir in 1998 when she publishes her research on clitoral anatomy in the *Journal of Urology*. Like Sevely and the authors of *A New View of a Woman's Body*, the Australian urologist demonstrates that the clitoris isn't a small button of tissue but rather an intricate erectile organ that extends from the visible crown deep into the body, where it's closely connected to the vagina and urethra.[85] Echoing Lisa Jean Moore and Adele E. Clarke, who deplore the "visual clitoridectomy" evidenced in medical textbooks, O'Connell argues that the clitoris is described in imprecise, incomplete, or incorrect terms, even in current publications on anatomy.[86] People have known about the complex internal structures of the clitoris since de Graaf's detailed studies in the seventeenth century, but this knowledge has routinely been forgotten or ignored. O'Connell doesn't draw on relevant past findings either, unlike Mary Jane Sherfey, Sevely, zur Nieden, or activists in the women's health movement. She does, however, accomplish something her predecessors struggled with: gaining widespread attention.

The doctor's research spreads across the globe, in medical journals and popular media alike.[87] Finally. O'Connell defines the clitoris as a system of erectile features and nerves.[88] Her research involves dissecting more than fifty clitorises, and her findings are clear: it makes no sense to differentiate between a clitoral and vaginal orgasm. Wherever the female genitals are stimulated, the entire clitoris is always involved.

An important conclusion that receives less attention, though, is the relationship between the distal part of the urethra, vagina, and clitoris. Easy as it is to separate these parts in a dissection, they are so closely connected (down to their nervous and circulatory systems), they probably shouldn't even be viewed as discrete features: "The distal urethra and vagina are intimately related structures, although they are not erectile in character. They form a tissue cluster with the clitoris. This cluster appears to be the locus of female sexual function and orgasm."[89] In a 2006 chat with the BBC, O'Connell reiterates the vagina's astonishing proximity to the clitoris: if you remove the skin on the side vaginal walls, the clitoral bulbs come into view.[90] She will later name this unit—comprised of the clitoris, distal vagina, and urethra—the "clitoral complex"—a term that evokes Sevely's "C. U. V. response." When the clitoris, urethra, or vagina is stimulated, the three respond as one, even if respective patterns of sexual response differ.[91]

Awareness today of the size and complexity of the clitoris is largely because of O'Connell, even if this knowledge hasn't exactly hit the mainstream. Still, she doesn't turn Queen's evidence in defense of the female prostate or ejaculation. She rules out the anatomical existence of the G-spot and barely mentions the paraurethral glands (avoiding the term "female prostate" altogether) or female ejaculation in her detailed reports.[92] Although she references other research—the fluid some women express from the urethral opening during intense sexual arousal, O'Connell notes, appears to originate in the paraurethral glands, demonstrates high PSA values, and is not urine—she does not add any of her own discoveries or reflections on the matter.[93]

A sexual technique practiced across Central Africa hinges entirely on the clitoral complex just as O'Connell and the women's movement depict it—as a sizable structure extending deep into the body. The method is known as *kunyaza*, and its sole objective is to drive women to the point of orgasm and expressive genital expulsions.

BRIEF ASIDE: GET THAT LADY A BUCKET! KUNYAZA IN CENTRAL AFRICA

> If your lover knows what he is doing, you'll pour rivers.
> —Nsekuye Bizimana, "Another Way for Lovemaking in Africa"

Known as kunyaza (literally "to make urinate") in Burundi and Rwanda, and *kachabali* in Uganda, this celebrated sexual technique for heterosexual encounters is deemed a surefire way to produce successive sodden orgasms in women.[94] Popular in Tanzania and the Democratic Republic of Congo as well, "wet sex"—a fun alternative to the "dry sex" practiced in sub-Saharan Africa—has been around for at least 150 years.[95] *Sacred Water* (2016), Olivier Jourdain's documentary film about kunyaza, relates the creation myth behind the technique. The queen is alone, thirsting for sex, and orders a slave to sleep with her. Terrified of penetrating the queen, the man lies trembling on top of her, and the vibrations of his body prompt her waters to flow. These very waters create Lake Kivu, after the queen sets out her sleeping mat to dry the following dawn.

Talking, tenderness, and time are key to kunyaza. The central objective is a woman's orgasm. Rwandan radio journalist and sexologist Dusabe Vestine mentions *gukuna* as part of kunyaza, with this term describing a manual, permanent elongation of the labia minora. "Water springs more easily if you have pulled the lips," she tells Jourdain. But does that water really exist? "White people don't believe in it because they don't know it," Vestine counters. Eighty to 90 percent of women in Rwanda produce this water, she says; it's an important part of their culture. "That's like the past. It is Stone Age," a young woman on the street laughs, in apparent contradiction. "Modern culture destroyed it," her male companion jokes.

In the late aughts, Nsekuye Bizimana, a Rwandan physician based in Berlin, publishes an article in *Sexologies*, the official journal of the European Federation of Sexology, and a book on the subject of kunyaza. Rather than emphasizing vaginal penetration, kunyaza inclines toward thorough stimulation of the vulva and deeper regions of the clitoris. The male partner rhythmically taps his erect penis against the labia minora, vaginal vestibule, clitoral hood and crown, urethral opening, vaginal opening, and perineum. The tapping gradually speeds up, interrupted by quick thrusts into the vagina. Kunyaza thus targets every erogenous genital area either directly or indirectly. According to Bizimana, kunyaza arouses women to the point of orgasm within three to five minutes, and often repeatedly. Women enjoy this technique, he explains, especially if vaginal penetration doesn't do it for them.

Many women produce a great deal of fluid during kunyaza—up to a quart or more of a thick, odorless liquid that might be clear or opaque. It flows from the urethra and/or vagina throughout the act, not just at climax. In interviews with Bizimana, men describe their penises being deluged by warm fluid. Bizimana theorizes that the ejaculate is a mix of genital juices, including fluids produced in the prostate and Bartholin's glands, vaginal secretions, and maybe even urine.

Women who ejaculate in abundance are known in Rwanda as *kingindobo* ("place a bucket under her") or *shami ryikivu* ("arm of Lake Kivu"). Such women are respected for their sexual prowess. Bizimana writes that there are individualized terms for all visible female genitalia; for instance, there's a different word for a prominent pubic mound than a flat one. The characteristic sound of kunyaza is described as "the dog lapping water." Women are expected to ejaculate, and men are expected to get them there. There's pressure on men in Rwanda to master kunyaza. In her chapter on Vestine, Swiss journalist Barbara Achermann quotes one young man's take: "You're a poor devil if you can't do it. Every woman will run away from you."[96]

Things get tricky for women if they fail to ejaculate during kunyaza "because [they aren't] viewed as 'real women.' Such women are a known quantity, referred to as *rwasubatre*, which means: sex with her is as hard as splitting granite. It's considered grounds for divorce or seeking a new partner if a man discovers he has married a *rwasubatre*." A woman's emissions during kunyaza, Bizimana says, signal to her lover that he has satisfied her. Wet sex, he continues, is a uniquely pleasurable and relaxing form of lovemaking for women: "Many women are so proud of ejaculating that they put their mattresses on full display to dry them out. Some women brag about how much fluid they express. There's another reason too: It makes us feel utterly relaxed and clean. After giving birth in particular, it feels like ejaculation cleans us from the inside, out."[97]

6

"IN CONTROL OF EJACULATION": SUPER HEROINES OF FEMALE EJACULATION

As we've seen, the Western discourse surrounding female ejaculation in the 1980s and 1990s is defined by specialists, doctors, sexologists, and sex therapists. That is, until three ladies take the stage who channel their strength and spirit, rage and sense of humor, in connecting coming with feminist practice and theory. Shannon Bell, Annie Sprinkle, and Deborah Sundahl are pioneers, an early generation of ejaculation superheroines. They aim to change the narrative by sharing their own experiences and demonstrating the phenomenon onstage and on-screen. These three North American activists fuel the rediscovery of female ejaculation and lead the charge in empowering women to rejoice in their juice.

"IT FEELS FANTASTIC"—SHANNON BELL

Bell, born in 1955, is a feminist writer, artist, and member of the Canadian queer community. She's also a poli-sci professor at York University in Toronto, where she teaches postmodern theory, fast feminism, cyberpolitics, and identity politics. Bell learns to come in the mid-1980s, prompting Sprinkle to nickname her "the Ejaculator."[1] From the first, Bell views this controlled efflux as a political statement. She calls for the reappropriation of a skill that has been unfairly reserved for just half of the population—men.

By ejaculating, Bell claims space for herself, marking her territory with a deeply (male) gendered gesture.[2] She views it as a biological function of female bodies and every woman's right. Bell perfects the act, a deliberate physical technique, and can make herself come in one to three minutes, up to fifteen times an hour![3] She ejaculates in response to clitoral or vaginal stimulation, sometimes in conjunction with orgasm, but not always. Bell, who is slim, can feel her prostate swell with fluid through her abdominal wall, and she has described how the color, odor, and texture of her ejaculate change over the course of her cycle. Sometimes she produces little fluid, and at other times her spray could knock down walls. One thing's for sure: she's a dead shot. Bell, Rebecca Chalker writes, is "renowned for her ability to ejaculate on a dime."[4] The artist herself states, "Female ejaculation is body power. It is not really about sex; it is about erotic power."[5] One step ahead of naysayers, Bell has sent her ejaculate to the lab; the fluid is distinct from urine.[6]

Personal experience is at the heart of Bell's writing, photography, and videos. In 1989, she publishes "The Everywoman's Guide to Ejaculation"—the first how-to of its kind—in the lesbian magazine *Rites*. That issue is released on International Women's Day. Readers enjoy black-and-white shots of Bell coming and learn about the history of female ejaculation. Later that year, Bell and Kathy Daymond coproduce a thirteen-minute short called *Nice Girls Don't Do It*, the first nonmedical film about female ejaculation. It lands somewhere between instructional video and porn, and in it Bell explains what she's doing as she masturbates and repeatedly comes, all while a pretty blond sits on her face. From this early work on, Bell articulates how important controlled coming is to her: "There is a big difference between ejaculating and

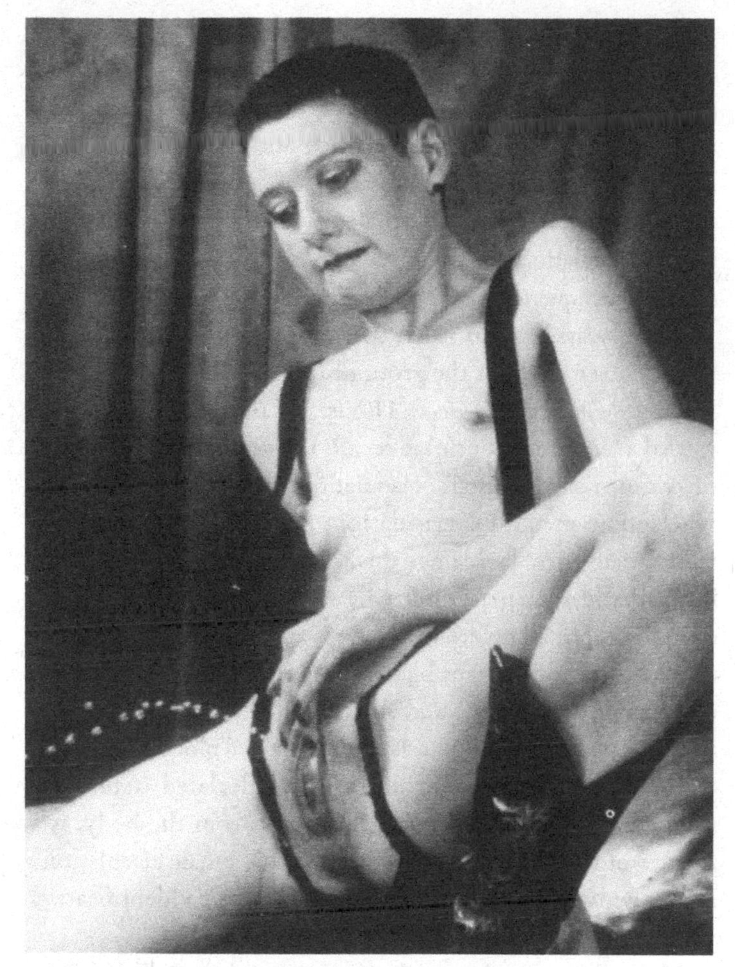

Figure 6.1
Shannon Bell, photographed by Annie Sprinkle.

being in control of ejaculation, ejaculating when you want to. A lot of women who ejaculate see it as something that happens, not something that they can take as their own activity and empower it and make happen."[7]

Ejaculation is enjoyable. "It feels fantastic."[8] It's raw and wild and pushes boundaries. The ejaculating body isn't maternal; it's an aggressive, self-defined, sexual body. Ejaculation represents nothing short of personal autonomy and physical self-actualization. A different form of femininity is manifesting here. According to Bell, male ejaculation is just a copy of female ejaculation. And a lousy one at that, considering such aspects as frequency or the volume of fluid expressed.

Three years later, in 1992, Bell teams up with Sundahl and Carol Queen to shoot the groundbreaking film *How to Female Ejaculate: Find Your G-Spot*. Thousands of women see it. "It caused a sensation!" Sprinkle recalls ten years later.[9] Hundreds of women attend Bell's ejaculation workshops. Feminists who ignore or dismiss ejaculation, rather than championing it, infuriate her. She blames this tendency on a form of feminism that casts female and male bodies as opposites; difference feminism aligns its worldview and gender politics with the "natural" differences between men and women. Women menstruate, give birth, and nurse, whereas men—and only men—ejaculate. Bodily fluids like menstrual blood, discharge, and milk, which flow or drip gently, are declared "feminine" fluids. Ejaculate, which spurts forcefully from the body, is a marker of the male body. The very term "female ejaculation" is problematic, critics argue, because it leads to identification with a male faculty.

Meanwhile, female ejaculation is central to Bell's concept of a "female phallus," the second component of which is the "internal erection" of the female prostate. Bell exhibits the stimulated tissue by inserting a speculum—an instrument used for vaginal examinations—and turning it to the side to bring the anterior vaginal wall and swollen G-spot into view. Bell believes in *one* human body and mutable boundaries between male and female, with men's and women's genitals corresponding in development, structure, and sexual response. The (re)discovery of the female prostate and ejaculation demonstrates that gender is a "cultural hallucination."[10] There's a reason the folks who challenge a binary, heteronormative world order—lesbians, gender terrorists, or gender deviants, as Bell calls them—are the ones leading the charge on ejaculation.

To Bell's mind, the ejaculating female body is the postmodern body par excellence, as it dismantles the gender binary and puts body equality on display: "To accept female ejaculate and female ejaculation one has to accept the sameness of male and female bodies."[11] The term "female ejaculation" signifies this equality and should remain in use, as it also helps redefine male ejaculation—which is, Bell maintains, the lesser of the two. Not only do women emit far more fluid, they can do it over and over again. Shouldn't that spot on the podium be reserved for women and their spurtuosity? In her 2010 book *Fast Feminism*, Bell recalls the Michigan Women's Festival, which hosted a tongue-in-cheek "Ejaculathon." Who can come quickest? (Two seconds to victory!) Who can spray farthest? (Almost twenty-three feet!) Who can furnish the most fluid, and how many times in a row? The communal event represents "body power, politics, skill, fun and the spectacle of female ejaculation."[12] It's a lighthearted celebration of a successfully reclaimed skill.

Many feminists in the 1980s and 1990s fear that medicine and mainstream pornography will seize on ejaculation, then misinterpret and standardize it, all in an effort to turn a profit. They have reason to worry. Bell describes a veritable "knowledge explosion" in sexology, esotericism, and adult entertainment between 1995 and 2005. The porn industry makes millions on (real and phony) depictions of female ejaculation and squirting, with hundreds of thousands of clips now available on YouTube, YouPorn, Pornhub, and Xtube. Such ubiquity has largely stripped female ejaculation of its subversive power, Bell acknowledges with dismay. She has called for consensual "postporn" that eschews heteronormative expectations and lenses, celebrating queerness that "disrupts and changes the actions and genders it depicts."[13]

Bell's interpretation of vulval ejaculation makes real demands. She wants equal rights and opportunities in keeping with the body equality epitomized by the act. Her argumentation does a double backflip in time, landing in the Middle Ages and antiquity by way of the nineteenth century: Bell is making political, cultural, and social demands based on a bodily skill. In the same way women in the nineteenth century were incapable of thought because their nerves were so weak, or the way menstruation was reason enough to bar them from work outside the home, women's ejaculation is now touted as proof of their sameness to men. Bell's concept of a female phallus is original and brave, and her facetious assertion that female ejaculation "relieves the phallus of its patriarchal burden" is a suitably acidic commentary on women's reappropriation of a sexual faculty men now have to share.[14]

"IT'S WORTH LEARNING"—ANNIE SPRINKLE

Sprinkle, née Ellen F. Steinberg in 1954, is a dazzling figure in US feminist performance art and LGBTQ movements.[15] She is an activist, healer, educator, sex researcher, and multimedia artist who identifies today as ecosexual. Her work challenges traditional representations of female bodies. She advocates for sexual education reform, decriminalizing and destigmatizing sex work, and women deciding how to view their own bodies.

Sprinkle fights for a sex-positive, loving society free of fear and repression. Her projects flirt freely with art, kitsch, pornography, education, esotericism, and politics. She challenges taboo topics, pushes boundaries, and uses humor to teach about sex, philosophy, spirituality, and what connects them. Sex, for Sprinkle, is a "path to enlightenment." Her pseudonym alone illustrates her love of wetness: "I like waterfalls, piss, vaginal fluid, sweat, cum—anything wet. I love rain, and I practically *grew up* in a swimming pool. So 'Annie Sprinkle' seemed perfect!"[16]

The moniker is not, however, a nod to vulval ejaculation. When Steinberg rechristens herself Annie Sprinkle, the young woman from Pennsylvania has never heard of the phenomenon. Sprinkle works as a prostitute, stripper, and dancer in 1970s' New York City. "I wanted to try everything with everyone," she recalls from her beginnings in the sex industry. She becomes an adult film star, and publishes articles and photos in porn magazines. She appears in more than a hundred adult films written, directed, and produced by men, until enough is enough: in 1981, Sprinkle shoots her very own mainstream porn, *Deep Inside Annie Sprinkle*. At the time, "when I made this movie, a lot of people, including many porn directors, weren't sure if women could actually have real orgasms," she writes. "And even if they could, they weren't important, anyways, because there was no sperm! But now I was directing. I wanted to show a real woman's orgasm."[17]

Sprinkle writes, directs, and stars in *Deep Inside Annie Sprinkle*. The film breaks rule after rule of popular pornography, only to become the second-best-selling adult film of 1982.[18] What a coup! And in hard-core cinema, of all places, the "last bastion of masculine, phallic discourse."[19] *Deep Inside Annie Sprinkle* subverts hard-core conventions. The filmmaker disrupts a space in which men have always looked at and into women. She returns the gaze. While men are accustomed to consuming female bodies visually, Sprinkle interrupts the hand jobs and humping to address the audience. She exhibits active agency in defying a traditional view of the female body, an anonymous and distanced view that nevertheless subjugates its object. She talks to viewers, inviting them to interact with her. The film also depicts ejaculation, although Sprinkle doesn't realize it at the time: "There was no name for what had occurred. I had no knowledge of what my body was doing."[20]

The film marks a turning point in Sprinkle's career and liberates her from "junk sex." No longer interested in "being other people's fantasy," Sprinkle turns political.[21] In 1985, in the midst of the sex wars—clashes within the feminist movement over such issues as sexuality, pornography, prostitution, and censorship—Sprinkle pivots to performance art, joining the cast of *Deep Inside Porn Stars*. This new self-image is further reflected in the "Post Porn Modernist Manifesto" (1989), which she coauthors with Veronica Vera, Frank Moores, Candida Royale, and Leigh Gates. Penned during the "Rubber Age," in the shadow of the AIDS crisis, and despite homophobia and censorship battles, the manifesto celebrates hetero-, homo-, and bisexual love as a life-giving, unifying, positive force:

> We embrace our genitals as part, not separate, from our spirits.
> We utilize sexually explicit words, pictures, performances to communicate our ideas and emotions.
> We denounce sexual censorship as anti-art and inhuman.
> We empower ourselves by this attitude of sex-positivism.
> And with this love of our sexual selves we have fun, heal the world and endure.[22]

In her lighthearted performances, Sprinkle invites audiences to acquaint themselves with the female body. Her "Public Cervix Announcement," from *Annie Sprinkle's Post-Porn Modernist* show, is the stuff of legend: the artist warmly encourages audience members to peer inside her to see the cervix. First, with the help of visual aids, she explains the uterus, vagina, ovaries, Fallopian tubes, and cervix. She drills viewers on terminology, then gets cleaned up, goes pee, and settles into a low armchair. Chatting with the audience, assuring them that neither the vagina nor the cervix has teeth, Sprinkle inserts a speculum and calls folks up onstage to view her "beautiful" cervix with the help of a flashlight. She greets those brave enough to step forward, making small talk as they look inside her body, asking them to comment on what they see. After all, it's important that everyone in the crowd understand what's between her thighs, even if they don't dare join her on onstage.

The "Public Cervix Announcement" creates an atmosphere that transcends voyeurism or exploitation. The verbal and visual exchange is tender and moving, German author Mithu M. Sanyal writes about Sprinkle's show at Düsseldorf's Kunstpalast Museum.[23] The cervix is the gateway to life itself and we don't even know what it looks like? Sprinkle is still active today, putting her cervix on display, and urging women to check theirs out too and get to know this important part of their bodies. The notion of reaching for the speculum to examine your own cervix and that of other women, determining where someone might be in their cycle based on its appearance, is nothing new: members of the women's health movement did the very same, which also yielded the photo series on the cervix in *A New View of a Woman's Body*. Now known as "vulva watching," the practice has made a real comeback, whether in groups or private sessions.

More than twenty-five thousand people have seen Sprinkle's cervix.[24] Though hers has been on display in more than twelve countries, the cervix remains enigmatic to Sprinkle: "The female body will always be a very great mystery, no matter how many you see or how much knowledge you achieve. You can never demystify a cervix."[25]

Sprinkle is one of millions of readers who learns about the G-spot and female ejaculation from Ladas, Whipple, and Perry's book in 1982. She later meets Perry in New York, and he shows her where her G-spot is. (At other times Sprinkle has said her friend Barbara Carrellas helped in the search.)[26] In retrospect, Sprinkle has expressed ambivalence about the consequences of *The G Spot*. Though it presented fascinating new discoveries and offered relief to women who feared they were incontinent, the book also created a new form of performance pressure in bed. Sprinkle recalls how threatened some men felt at having been robbed of something they'd thought exclusively theirs. Women who failed to come (or men who failed to get their partner to come) fretted over their apparent incompetence. Lesbians long in the know about the G-spot were peeved because they didn't profit from the book's recognition and financial success.[27] After all, it was a group of lesbian activists that first informed Whipple and Perry of this vaginal zone.[28] Feminists were upset that, of all the terms they could have chosen, the authors named an intimate female body part after a man (Ernst Gräfenberg), and after discovering their G-spot, some women were disappointed because it didn't lead to superorgasm. Nevertheless, many women were eager to experience this long-lost sexual response, and learn and lay claim to it in an act of empowerment, as Sprinkle would say.[29]

Ten years after *The G Spot* was published, an equally pioneering film is released, Sundahl's *How to Female Ejaculate: Find Your G-Spot*. Thousands of women watch the instructional video, which also depicts women coming, among them Bell. As Sprinkle exclaims in her foreword to Sundahl's book *Female Ejaculation and the G-Spot*, "The world has never been the same!" These days, "women (and other genders) are freely enjoying big, woman-made, wet spots in beds around the world," with women finally convinced their experiences are real. Things looked different in the early 1990s: "How exciting it has been to have witnessed so much growth in the world of female sexuality in such a relatively short time," Sprinkle reflects.[30]

In her own work, Sprinkle increases ejaculation's visibility. She writes and teaches about it. She masturbates and comes onstage and on-screen. One of her biggest films is *The Sluts and Goddesses Video Workshop, or How to Be a Sex Goddess in 101 Easy Steps* (1992). Sprinkle writes and stars in the fifty-two-minute movie, codirected by Maria Beatty, which includes an orgasm lasting five minutes and ten seconds (as measured by a sort of galvanometer) and an ejaculation scene caught on two cameras.

Sprinkle's public persona and projects radiate a spirit of generosity and ease. She doesn't lose sleep over vulval ejaculate, and doesn't insist on differentiating between come and urine. The artist recognizes several erotic genital fluids: vaginal secretions, the spray or dribble of liquid from the urethral opening that isn't pee, *and* sexually charged urination. The "golden shower," which Sprinkle defines as "the art of erotic urination during sex-play," can lead to ejaculation and vice versa. She resists the rigid definition and separation of these fluids in the same way she plays with such false dichotomies as man and woman, queer and straight, active and passive, goddess and slut, object and subject. Whether pleasure gives way to ejaculation or urination—why does it even matter, what's being sprayed?[31]

In her book *Annie Sprinkle: Post-Porn Modernist* (1991), Sprinkle dreams up a utopian vision of the world, in which motley, easeful, consensual sex and vulval ejaculation come standard:

> I have a vision for the future, of a world where all the necessary sex education will be available to everyone, thus, there will be no more sexually transmitted diseases. . . . Fetish lingerie and sex toys will be freely distributed to all people. People will be able to make love without touching if they choose. Men will be able to have multiple orgasms without ejaculation, so that they can maintain erections for as long as they want. Women will ejaculate. It will be possible to make love anywhere in public, and it will not be impolite to watch.[32]

Sexuality and spirituality go hand in hand for Sprinkle. Sex heals. It connects individuals with one another, the earth, and the universe. Sex teaches us the secrets of life and death. A sexually experienced woman is a "divine and extremely powerful force that . . . has the potential to contribute to the well-being of all life on earth."[33] Sprinkle no longer dedicates projects to ejaculation, something she now considers mainstream, though it remains a major aspect of sexual potential in women and vulva-havers.

These days, Sprinkle lives as an ecosexual who prefers the E(co)-spot to the G-spot. Her teaching now focuses on a love of trees, rocks, and plants. Nature is no longer your mother; it's your lover. "We make love with the Earth through our senses," Sprinkle and wife Beth Stephens declare in their "Ecosex Manifesto."[34] The couple presents their vision at documenta 14 in 2017, leading Ecosex Walking Tours around the city of Kassel. Together with Stephens and Jennie Klein, Sprinkle outlines her ecosexual evolution and vision in *Assuming the Ecosexual Position: The Earth as Lover* (2021).[35] Sprinkle has emerged as a woman united in love with the natural world: "We caress rocks, are pleasured by waterfalls, and admire the Earth's curves often."[36]

"A WILD, UNRESTRAINED EXPERIENCE"—DEBORAH SUNDAHL

The third queen of ejaculation heralding from the United States is Sundahl, an author and sex educator.[37] Sundahl (1954–2023) is a prominent figure in the women's and lesbian movements in California from the 1980s onward. Until her death, she remains a foremost authority and best-selling author-cum-filmmaker on the topics of female ejaculation and the G-spot. Sundahl founds the Female Ejaculation Sex Education Institute, and runs workshops in North America and Europe, offers interactive online seminars, and lectures around the world.

Sundahl moves to San Francisco as a young student in the early 1980s—an arrival that is at once a point of departure—and becomes a radical feminist. Sundahl fights the patriarchy, has her first lesbian relationship, and discovers "sexual feelings I did not know my body was capable of feeling. That kicked open the door to explore personal exploration of women's sexuality, my sexuality."[38]

In 1984, Sundahl and Myrna Elana found *On Our Backs* (*OOB*), subtitled *Entertainment for the Adventurous Lesbian*, the first erotica magazine for lesbians, by lesbians. The publication is revolutionary. Lesbians and lesbian sexuality are depicted in unprecedented ways, with color photographs to match. The magazine features queer women of all stripes—femme and butch, women of color, punks, Asian dykes, and S&M and leather lesbians, bucking any notions of sapphic sex as necessarily vanilla.

OOB is a commercial publication with nationwide distribution, started at the height of the feminist sex wars. The launch is an act of empowerment. With it, Sundahl and her cofounders make a pledge to personal autonomy and pleasure. It's also a political statement. In an editorial in the first issue, they write, "Lesbian sexuality . . . is as diverse and multiflavored as all the parts of our lives. Towards the goals of sexual freedom, respect and empowerment for lesbians, we offer On Our Backs."[39]

Not only does the magazine nix common images of lesbians, it takes on such taboo matters as AIDS, the enjoyment of gay porn, anal sex, and transsexuality. *OOB* covers the G-spot (closely tied to the contentious subject of vaginal penetration), and paves the way for a more easeful approach to vibrator and dildo use. The magazine is strident, raw, and explicit. For most of the feminist and lesbian scene, it's also too sexy. Women's bookstores refuse to sell it; academics and intellectuals keep their distance as well. Around the same time, Sundahl and Nan Kinney start Fatale Video, a film production company for lesbian porn and how-tos that will eventually put out some of the most successful videos on female ejaculation.

Sundahl has been curious about the G-spot and ejaculation since 1984, when—to her own amazement—she gushed all over her hardwood floor during sex. After mopping up the puddle, she started digging. Her exploration, which includes many conversations with fellow ejaculators, leads to *How to Female Ejaculate: Find Your G Spot* (1992), which reaches an audience of thousands. The video is a milestone, a fun film made for women, by women, intended to instruct and grant them a spectacularly new perspective on their own bodies. Sundahl puts her G-spot on display too, turning the speculum to reveal the anterior vaginal wall: she flashes her female phallus, as Bell might say. Sundahl's film diverges from the sterility of *Orgasmic Expulsions of Fluid in the Sexually Stimulated Female* (1981), for instance, a nine-minute educational video in which two women lie on a gynecological examination table, their heads out of frame, while a doctor's gloved hand stimulates their vaginas. It's a troubling production that objectifies the female body, diametrically opposed to the joyful performances put on by Bell, Sprinkle, or Sundahl.

How to Female Ejaculate: Find Your G-Spot (Fatale Video) is a hit, following such resolutely nonmedical films as Bell's *Nice Girls Don't Do It* (1990, directed by Kathy Daymond) and *The Magic of Female Ejaculation* (1992), by Daymond and Dorrie Lane.[40] Sundahl goes on to produce a series of successful ejaculation videos that are still for sale: *Tantric Journey to Female Ejaculation: Awaken Your G Spot* (1998), *Female Ejaculation for Couples: Share Your G Spot* (2003), *Female Ejaculation and the G Spot: A Lover's Guide* (2006), and *Female Ejaculation: The Workshop* (2008).

Sundahl also writes a book on her life's work. *Female Ejaculation and the G-Spot* (2003) covers all the bases: it outlines the history of female emissions, explains the anatomy, and provides exercises to strengthen the pelvic floor muscles and "heal" the G-spot. It also shows women how to practice ejaculating, gives tips for their partners, and explores the spiritual side of the "feminine fountain." It's a good read, though at times Sundahl can oversimplify things, provide fuzzy "facts," or forgo substantiating her arguments.

Sundahl's interpretation of female ejaculation deviates significantly from Bell's. "Inherently feminine, . . . fun and sexy," female ejaculation is "every woman's birthright," Sundahl believes. She links it to a primal femininity. Ejaculation "can feel strong, explosive, tumultuous, passionate, large, messy, chaotic, complicated, and curious." It also defies constraints applied mostly to women during sex that stop them from going wild, experiencing ecstasy, and relinquishing control. To Sundahl's mind, ejaculation means banishing the corset of tidy, dry femininity as well: "If we let go, we'll gush all over our partner's face, chest, thighs, cock, and the bed and pillows, too! Not very ladylike, is it?"[41]

Sundahl informs readers that the G-spot and female prostate are one and the same.[42] All women have a prostate that produces ejaculate, which means all women can learn how to ejaculate from the urethra. "When we're familiar with the anatomy of our genitals, ejaculation is just a by-product. And how do I know that? Because I don't just theorize about it; I work with real, live women!" Sundahl expresses in a 2015 interview.[43] She contends that an orgasm driven by G-spot stimulation is very different from a clitoral orgasm. The former "opens your heart" and has "far more to offer."[44] She distinguishes between clitoral and G-spot orgasms as well as the uterine orgasm. This third form is unleashed by "deep and rapid thrusts that jostle the cervix."[45]

Many women criticize the notion that orgasms can be classified and differentiated according to their source. Sundahl's claim that vaginal orgasm is superior to clitoral orgasm is also problematic. The classification of orgasms (better, deeper, more spiritual, more cleansing, more healing, etc.) is a counterproductive, if lucrative, practice. Just look at Sigmund Freud, feminists like Alice Schwarzer, or personal trainers who charge up the wazoo to help clients unlock new levels of sexual release. The current tally of orgasmic types is up to twelve, including the "zone orgasm": "The zone orgasm differs from person to person. As its name implies, any number of areas might lead to orgasm when stimulated."[46]

Rankings like these impede people's relationships with their own bodies. In the 1990s, Sundahl discovers Tantra, and begins to incorporate Tantric philosophy and practice in her work: "To me, learning about the Tantric path has been critical to understanding the role of the G-spot in women's emotional, physical, and spiritual lives."[47]

BRIEF ASIDE: FEMALE EJACULATION IN TANTRA, PART II

Tantric philosophy and physical practices gain traction in the United States and western Europe in the 1960s and 1970s, albeit in an abridged form tailored to the times. Neo-Tantric guru Nik Douglas declares it an "engine of political change."[48] Tantric sex and eroticism are part of a societal push in the 1970s for increased political, religious, moral, and sexual freedoms—an ideal canvas for notions of natural, liberated, equitable sexuality. As it works its way into everyday culture, Tantra is adapted to meet Western interests and needs. In the worst case, it's sold as an esoteric sex technique to be mastered in a few easy steps, provided one selects the right guru, book, or workshop. California-based Indologist David Gordon White decries the trend, quipping that New Age Tantra is to medieval Tantra as fingerpainting is to real art.[49]

Bhagwan Shree Rajneesh, who would become a notorious cultist also known as Osho, popularizes his own version of neo-Tantrism after relocating to New Jersey from India in the early 1980s. Osho links spiritualism to sexuality, and promotes self-discovery and self-realization as part of one's sexual experiences. Neo-Tantra draws from Wilhelm Reich's teachings, integrating yoga and breath, body psychotherapy, and psychology. Tantra is seen as historically having held women and the feminine in high esteem. Bodies and sexuality are experienced in a new way, freed from societal shackles, which is what makes the practice so attractive and suitable to emancipatory movements of the time. Men confront their emotions and bodies in Tantra, while women work on sexual self-confidence and discover new emotional and erotic independence. The goal for both is to achieve "holistic" personhood melding body and mind, spirituality and sex. Tantra provides men and women the opportunity to interact in an intimate, consensual, voluntary, equitable, and mutually beneficial way, writes author and esoteric practitioner Rufus Camphausen. In his view, this novel expression of the male-female relationship differs drastically from the "war of the sexes."[50] These days, Tantra is just another lifestyle option in Western (spiritual) consumer society. The G-spot and female ejaculation—ejaculate in Tantra is often referred to as *amrita*—are largely associated with Tantra in North America and Europe. Those who wish to confront personal trauma and unleash hidden energies, so the thinking goes, must be ready to confront the G-spot and ejaculation.

Bell called on women to reclaim their right to ejaculate, asserting (by squirting) their due, space, and freedom. Sprinkle presented ejaculation as yet another miraculous manifestation of the goddess's fluificent form. And now vulval ejaculation is something to be attained. In Tantra (and I exaggerate), a woman must discover her G-spot and experience ejaculation before she's considered a fully realized, orgasmic person. Caroline and Charles Muir, US Tantra pioneers of the

1970s, assign ejaculation to the seventh and highest plane of female orgasm. When women find their "sacred spot"—the G-spot—and free it of "painful experiences," they will have multiple protracted orgasms, repeatedly spraying the "nectar of the goddess."[51] The Muirs are certain all women have the potential to ejaculate, but can neither practice nor otherwise influence the way they spray. Bell's idea of women squirting whenever, wherever, and however often they want has turned into "I am ejaculated." When a woman ejaculates, "it's like a divine gift."[52]

Sundahl considers the G-spot a "gateway to deeper aspects of sexual expression and intimacy." Traumatic experiences are stored in the G-spot and pelvic floor. Confronting this zone helps break down its armor, heal emotional and sexual wounds, gain access to buried emotions, purify one's sexual energy, and gain greater or deeper health and spirituality. "Sexual intercourse with a pulsating G-spot and flowing ejaculate can open the heart to connecting with your partner on a deeper level of intimacy, love, and caring—moving toward the reverence for each other embodied by sacred sex."[53]

Italian Tantra practitioners Elmar and Michaela Zadra cast the G-spot as the "most erotic entrance to the unconscious realm," an important center of femininity for all to discover and awaken. By transcending "all manner of linear-intellectual approaches," the G-spot allows for "practical access to higher levels of consciousness." If a woman wishes to experience "deep" sexuality, she need only find and "unveil" her G-spot. Like Sundahl, the Zadras promise the sun and moon to those who conquer their G-spot, though this time there's a catch: for some women, discovering a deeper level of sexuality can be a nightmare. Those who decide to work with their G-spot must be prepared to confront the traumatic experiences stored there. The area falls within the first chakra, our sexual center: "When we liberate and live this energy, it thanks us with unbridled strength, an energy that was largely lost over centuries of women's sexual oppression." Vaginal orgasm, triggered by the G-spot, is a "journey to the deepest depths, to [a woman's] very own *terra incognita*," an implosion that opens her up to spiritual and transpersonal experience.[54]

What do the Zadras have to say about female ejaculation? Though the counterpart to male ejaculation, in women it's taboo, a "phenomenon with a thousand faces." And though known and "poetically" described in other cultures, we lack information and fitting terms for it today; beyond a vague image of "wetness" and the clinical term "female ejaculation," we are literally struck dumb by coming women. In a non-representative survey, the Zadras conclude that "about one third" of the sixty-five female respondents have experienced ejaculation.[55] In order to ejaculate, they write, one must be able to relax while in a state of heightened arousal. It can't be forced, nor is it required for a fulfilling sex life. It is, however, often accompanied by a feeling of deep connection to one's partner. For those uncertain as to whether they're ejaculating or peeing, the authors offer this pragmatic solution: eat some asparagus, wait until the urine has taken on the telltale odor of asparagine, and then compare the two fluids. Reminiscent of the early DIY days of the women's movement, it's a wonderfully tangible suggestion that brings people closer to their own emissions.

The Zadras write that, since the late 1990s, women have been devoting more time to their emotional and anatomical inner lives. They're focusing on feelings "in bed as well" and embracing their "passive side": "A receptive woman who allows space for her emotions" is back in demand.[56] It's no coincidence that women are now rediscovering the G-spot (although its heyday really was in the 1980s, when Ladas, Whipple, and Perry's book came out). An ejaculating woman, by the Zadras' account, is a hyperfeminine one. This woman, while bound emotionally to her partner, remains centered in herself, uncovering her passive side and ejaculating from a place of profound relaxation. In this rendering, ejaculation becomes an aspect of femininity diametrically opposed to the image Bell creates. It's impressive to see just how differently female coming can be interpreted and ideologically charged.

Today a range of Tantric institutes, teachers, and massage therapists provide resources on female ejaculation as well as offer workshops on yoni and G-spot massage. According to an association of German Tantric massage therapists, this form of massage is becoming "more and more socially acceptable." During a Tantric massage, some women will experience "a deeper and more complex dimension of their sexuality for the first time, whether in the form of vaginal orgasm, whole-body orgasm, or female ejaculation."[57] In a 2013 study, however, only 3.1 percent of female respondents state that "'training for the ability to ejaculate'—an established phenomenon in Tantric practice"—is how they picked up the skill.[58] Several critics lambaste the spiritual stranglehold on female ejaculation and fear its commercial exploitation.[59] The "McDonaldization of occultism," as Swiss writer Peter-Robert König calls it, could easily apply to Tantra. König jokes that, before too long, "frozen Amrita" will be available for purchase online.[60] With the proliferation of squirting videos, though, it's the porn industry making the biggest bucks—in the billions—off female juices.

SQUIRTING QUEENS: THE JIZZ BIZ MEETS THE BIRDS AND THE BEES

Feminist films and how-to videos for women, by women, help popularize female ejaculation starting in the late 1980s.[61] Producers like Sundahl and Kinney make entertaining films defined by empathy, integrity, and emancipatory drive. They aim to inform by depicting facets of female sexuality and pleasure that go well beyond old-school patriarchal representations.[62] These are the roots of the feminist pornography that takes off in the 1990s, so there's a certain irony—or perhaps tragedy—to the hype surrounding female ejaculation and squirting in hard-core porn, of all places, at the turn of the century.

Pornography is a billion-dollar industry these days. It's available worldwide and can usually be accessed for free. Female squirting or gushing (as it's known in pornspeak) is so popular that internet platforms like YouPorn, Pornhub, or Brazzers group these videos in their own category. In November 2017, Pornhub—the world's largest porn site, with seventy-five million visitors daily—reports a spike in popularity of squirting content between 2013 and 2015, the category now in the site's top twenty. What's more, "women are 44% more likely to search for squirting videos compared to men."[63] Users between the ages of eighteen and twenty-four are especially keen, while those sixty-five and older prove less curious about squirting.

Every year, more and more women are consuming hard-core porn. This burgeoning group is being courted by providers, who flag films as "woman friendly" or "popular with women." Women also use squirting videos to learn about female ejaculation and how to spice things up in bed. In

Vagina: A New Biography (2012), Naomi Wolf speaks with the manager at Babeland, a woman-friendly sex shop in Brooklyn, who says products designed to stimulate the G-spot are "flying off the shelves" because "pornography [has] begun to focus intensely on female ejaculation, leading women to wish to explore stimulation of their G-spots."[64]

Berlin-based sociologist, linguist, and PornYes activist Laura Méritt can attest to pornography's educational remit. When it comes to female ejaculation, she argues, today's major porn sites are "more progressive than any biology textbook."[65] Canadian physician and best-selling author Sharon Moalem also bemoans medicine's tepid interest in female ejaculation. He too points out that porn has taken over from science in teaching people about this natural aspect of female sexuality.[66]

Hard-core pornography has always been on the lookout for ways to show female desire and orgasm. Ironically enough, given the "maximum visibility" of their bodies in porn, women's arousal and climax are nigh impossible to capture.[67] Dirty talk and sounds, flushed cheeks, dewy skin, and parted lips serve as paltry evidence of real excitement. Though female ejaculation (or male, for that matter) is not the same as orgasm, it is now touted as "patent visual" proof of true arousal.[68] If a woman gushes, she has obviously climaxed. The money or cum shot—traditionally the moment in a pornographic film in which a male actor ejaculates—now applies to actors with vulvas as well.[69] Adult film stars like Hotaru Akane, Charley Chase, Annie Cruz, Jamie Lynn, Missy Monroe, Jenna Presley, or the nonbinary major star of queer porn Jiz Lee have gained notoriety for their squirting skills.[70] Their ultimate success, however, is all thanks to the goddess of gush herself, Cytherea, who in the early 2000s elevates squirting from a niche subset of porn to a mainstream standard.

Cytherea—née Cassieardolla Elaine Story—specializes in squirting scenes. Over the course of her brief career, the US porn star sparks a huge trend: "Before me, everything was about anal. But I literally changed the industry, from anal to squirting," she recalls in a 2014 interview about her comeback.[71] Cytherea is a dead shot and known for her ample spray. Others try to emulate her. Cytherea sees ejaculation and squirting as a skill people can pick up and practice. (She will later work as an ejaculation coach as well.) Her squirting videos are so successful, she explains, because real orgasms, real coming, and real female pleasure are on display. The squirt queen doesn't view coming as a feminist or emancipatory feat. Doing so in a man's face, however, *is* a culturally charged act: "The only time I really felt like an empowerment is because a lot of women can't squirt on guys and guys, they come on your face all day long, and they have the audacity to dog mine? I don't dog your come, don't you dog mine, asshole. So I kinda find it exciting that I will find a way to hit them in the face one way or another." Cytherea plans carefully for shoots, which are taxing: "It takes a lot out of my body. I need to prepare my body to squirt on film."[72]

Flower Tucci is another successful squirter. She challenges rival Cytherea to circus-like squirting competitions on Playboy TV: Tucci takes aim at a flowerpot while Cytherea tries to flip a light switch with her stream. Tucci has been a porn actor since 2002 and won several Adult Video News Awards—the Oscars of the US porn industry—including one for a squirting scene in *Squirt Shower 2*. She can squirt an average distance of 10–13 feet, with her personal best measuring 16.01 feet. In one interview, she recounts how learning to ejaculate took months of practice with vibrators and Kegels to strengthen her pelvic floor muscles: "Now if I have a clitoral orgasm, it's almost like I can't control it. It's so powerful that it'll just spray out, as if I put my finger on a water nozzle. The buildup feels like a discomfort, almost a burning, and then when I release it, I just want to take a deep breath and collapse. It's an overwhelming release of energy. I'm so satisfied, because not only have I clitorally come, but my body's released all the juice and the energy that it was building up."[73]

Both Cytherea and Tucci are pretty impervious to accusations of urinating: "I'm not a doctor, and I can't say what it medically is, except that it's clear and odorless. It's released from the urethra, but the clitoris, the urethra and the G-spot are all intertwined. My G-spot actually feels like a sponge." Champion squirter Tucci is certain all women can learn how to ejaculate. It's an "enjoyable part of a woman's sex life," something to be proud of: "Though, as a woman, I think you make yourself squirt. I want women to know that they are all capable of doing it. . . . It is a natural phenomenon and I wish that people were more accepting of women's sexuality and the release we can achieve that way. I love to see women squirt."[74]

Some squirting films center exclusively on female appetite and juices. In a group sex scene in *Swallow My Squirt, Volume 4* (2006), for instance, two ladies push the male costar to fuck their friend until she orgasms and squirts. His sole purpose is to screw all three of them exactly how they tell him to. The goal isn't male pleasure; it's a gushing orgasm. The film celebrates female orgasmic expulsions, just about obscuring the male experience. Squirting invalidates men's exclusive claim to ejaculation, an act so often used to symbolize women's submission to male sexual drive. The female cum shot dismantles the male "monopoly on pleasure and power in pornography."[75]

Be that as it may, commercial squirting upsets people, from porn opponents to feminists like Sundahl. Although a former adult film producer herself, Sundahl dislikes the way female ejaculation and squirting have been co-opted by mainstream straight porn. On the one hand, she admits that squirting videos have helped popularize the act, but on the other, they put a tremendous amount of pressure on women and paint an unrealistic picture of female coming. Women now "need" to produce huge amounts of fluid either because they demand it of their bodies or because their partners saw it in a porn: "Recently women have been telling me their partners expect that of them in bed. . . . My seminars are not about learning some circus trick. They're about the erotic body."[76] Besides, on-screen squirting is often faked, with actors emitting a fluid that's been pumped into their vaginas beforehand; either that, or they really are peeing.[77] People are starting to think female ejaculation is a myth again, and Sundahl blames the movies.

Censors in some countries classify female ejaculation and squirting as urination. In 2004, the British Board of Film Classification (BBFC) bans squirting in pornographic DVDs produced in the United Kingdom. Ten years later, the Audiovisual Media Services Regulations are updated to include streaming and video-on-demand services. Fisting, face-sitting, and spanking are also banned. The BBFC justifies its decision on female ejaculation by claiming these scenes depict urolagnia or "water sports"—sexual excitement derived from the sight or thought of urine—which is prohibited under the United Kingdom's Obscene Publications Act. In short, the BBFC questions the very existence of vulval ejaculation. Something similar happens Down Under, where the Australian Classification Board censors porn sites and films with squirting scenes, which it claims depict urination. As in the United Kingdom, the governing body states that "the fetish of 'golden showers' will be refused classification."[78]

These censorship cases inspire protests in England and Australia. As erotica writer Kristina Lloyd opines in the *Guardian*, "In refusing to accept the existence of female ejaculation, the BBFC positions itself as a shaper of female sexuality rather than a classifier of films."[79] Despite her own misgivings about mainstream pornography, Sundahl joins the debate, stating there's no excuse for the BBFC's "ignorance."[80] In her *Slate* column *The XX Factor*, author Hanna Rosin (*The End of Men*) ironically laments that only male ejaculation is allowed onscreen "because, well, women can't be equal in *everything*."[81] Feminist adult film director, artist, and politician Anna Span (née Anna Arrowsmith) is the first to fight the BBFC, with Sundahl's help. After Span's 2009 DVD *Women Love Porn* is censored because of a squirting scene, she and Sundahl present scientific evidence of the female actor's ability to ejaculate.[82] The filmmaker also pushes for a hearing before the Video Appeals Committee. The BBFC eventually green-lights the film, though it stops short of granting that female ejaculation exists. The board concedes that the film is not in violation of the Obscene Publications Act, as it features "so little focus on urolagnia."[83] Span and her production company Easy on the Eye celebrate what they deem a "historic victory."[84]

US production companies have long feared censorship as well. Tucci comments in 2011 that companies "don't want me to squirt at all, because everybody's getting sued for it. . . . I haven't shot a squirting movie in over one year. I'm an advocate for it, so I hate the government saying it's OK for these guys to bukkake girls, but god forbid a woman has a squirting orgasm on film."[85] Will legislation see to it that vulval juices vanish again from pornography? By 2014, when Cytherea returns to porn after a years-long hiatus, the tide has turned: what was once a prosquirting industry is now anti. She shoots just a couple more films with squirting scenes. The squirting boom of the aughts is over.

EPILOGUE

> The question isn't if female ejaculation is real. It's why you don't trust women to tell you.
>
> —Lux Alptraum, "The Question Isn't If Female Ejaculation Is Real," *Guardian*

> The eveningdress girl lifts her pelvis and cries out: from her shaved pubis flows a transparent stream that doesn't look like urine.
>
> —Virginie Despentes, *Apocalypse Baby*

It's a spring day in Berlin's Kreuzberg neighborhood. Laura Méritt greets me on the top floor of the prewar apartment building where she lives and works. Méritt is a sex activist, feminist, linguist, vulval ejaculation expert, and cofounder of the PornYes movement who also runs Europe's oldest feminist sex shop, Sexclusivitäten (Sexclusivities). For almost thirty years, Méritt has advocated for sex positivity and education as well as body diversity and autonomy. For instance, Méritt and her cohort launched the Million Puzzies Project to push back against "extensive advertising and the provision of seemingly 'objective' information [intended to set] new standards for the 'right' form, size and functionality of female sexual organs."

These Berliners celebrate vulval diversity. The group's "Pussy Profile" questionnaire asks online visitors what they call their genitals, how they style their pubic hair, and the size of inner and outer labia. The wealth of vulvas out there demands documentation because "the fact is there is no general norm. All of our bodies are different and unique. . . . The diversity of our bodies is valuable."[1]

Méritt has been active in the sex workers' rights movement since the 1980s, organizing the first annual observance of International Whores' Day in Germany in 1989. She traveled the country as a dildo dealer at a time when feminists still decried penetration, and in 1993 founded Club Rosa, a lesbian escort service. Méritt has always been comfortable in the gray areas of sex and politics that many other feminists daren't touch, like pornography, BDSM, and the spectrums of gender and orientation. For years she has campaigned for the visibility of female ejaculation. Today she prefers the term "vulval ejaculation," thus indicating that individuals with vulvas who don't identify as women naturally ejaculate as well.

An educator through and through, Méritt was the first to invite Deborah Sundahl to host workshops and lectures in Berlin more than eighteen years ago. The feminine fountain is a recurring topic at Méritt's weekly salon, where discussion ranges freely between sexuality, culture, gender, and (body) politics. She teaches classes on the function and anatomy of the female prostate for women, couples, and trans people who wish to learn more about squirting and how to do it. The workshops, titled "Female* Ejaculation and the G-Spot: We Squirt Back!" include a hands-on portion: disrobe, examine your vulva and G-spot (and maybe those of other attendees), spread out a towel, and get cracking.[2] While there's no "squirt guarantee," about half the participants in these workshops will come—and before an audience, no less. All genders can (learn to) ejaculate, Méritt contends. Then it's come one, come all.

Méritt has also commented on the "insane pressure" mainstream pornography puts on women—and men, she adds—to ejaculate and squirt. Everyone's expected to gush more and spray farther, but coming is about so much more than that, she says. It's about cultivating a positive approach to your own body, getting to know your physical self and its desires, and tending to those desires. "It's not about better, deeper, longer. It's about knowing what's possible, then deciding for yourself." Méritt's litany goes on. There is little mainstream knowledge of the female prostate's belonging to the seat of sexual potential (clitoris, pelvic floor, and urethra). Medical texts and sex ed resources are still diminishing the size of the clitoris. The female prostate is still going unmentioned. No one's teaching girls that breasts and labia can be different sizes; these young women are still ashamed of their genitals not looking "right."[3]

To address these issues, Méritt has revised and reissued the German version of *A New View of a Woman's Body*, the seminal publication of the women's health movement. Like Annie Sprinkle, Méritt models a breezy stance toward female sexual fluids. Whether prostatic secretions, squirting juices, or urine, "all fluids are great." The biggest challenge is "to establish a positive approach to our own bodies. Simply confronting the fear of its actually being urine—what's so bad about urine, anyway?" She calls categorizing orgasms "nonsense" and rejects squirting competitions. Méritt, who has a PhD in philosophy, wrote her dissertation on female laughter. With a chortle, of course, she shares one of her theses in parting: "I've always thought of laughter as a type of squirting."[4] Perhaps we should see squirting as a type of laughter?

Seventy years ago, US ethnologist Margaret Mead outlined the social conditions required to allow women to experience sexual pleasure. Society must place value on female desire. It must enable women to comprehend the mechanics of their sexual anatomy, and it must give them the skills to orgasm.[5] We still haven't managed to check these three boxes. It's bizarre that even now, in the twenty-first century, research into female sexual organs and physiology remains fragmentary. With a few exceptions—namely pharmaceutical moneymakers like "Viagra for women"—research into female anatomy and "psyche with regard to sexuality [also remains] woefully underfinanced."[6]

It comes as no surprise that almost every scientific study of vulval juices concludes with the recommendation for further research. How and what exactly do women and people with vulvas ejaculate? Does the fluid have antimicrobial properties?[7] Is there a link between squirting and UTIs? How should prostate carcinomas or adenomas be treated in women?[8] Does vulval ejaculate transmit HIV? Does the fluid play any role in conception? Is ejaculation easier after childbirth? Might prostatic secretions actually ease labor?[9] What is the physical mechanism underlying squirting, how does it work, and what *is* that fluid anyway? Further research is required. What research *has* been done, though, ought to serve as a benchmark. The female prostate must be acknowledged as a functioning organ. Ejaculation and squirting must be acknowledged as aspects of sexuality in women and vulva-havers.[10] Things we've known for decades about the prostate and expulsive sexual fluids should be taught in universities, schools, and health centers. As we've seen with the clitoris, it takes far too long for essential facts to reach the public—whether by means of education, publications, or state resources.[11]

Fortunately, more and more books on anatomy, gynecology, and urology are starting to depict the clitoris as a complex erectile organ. In 2016, French engineer and cognitive scientist Odile Fillod creates a 3D-printed model of the clitoris, complete with crura and erectile bulbs, and publishes the open-source code online: free clit printouts for all![12] Swiss biologist Daniel Haag-Wackernagel releases two further anatomical models in 2020. Surprised by "the dearth of decent images, let alone anatomically correct models of the clitoris," he develops detailed 3D renderings of the organ.[13] And in 2022, the revised sixth edition of the *Prometheus LernAtlas der Anatomie* is released. This reference work is the standard for anatomy instruction in medical programs across the German-speaking world. The new edition presents the clitoris (or "bulboclitoral organ") in its full complexity, including location and innervation. This is nothing short of a milestone. The online version of the *Thieme Atlas of Anatomy*, as the textbook is published in English, is updated in 2023. The detailed illustrations and descriptions are publicly available, without a paywall.[14] US clitoral education activist Jessica Pin celebrates the change, posting on Instagram, "We are making great progress."[15]

Popular science publications tend to describe some, though rarely all, parts of the clitoris—crown, hood, shaft, body, crura, and bulbs. And yet books that shrink the clitoris to the size of a raisin are still hitting shelves. Things look even worse for the female prostate, largely neglected by specialist and popular literature alike. Current medical guides tend to overlook the organ, and rarely mention female ejaculation and squirting.[16] Petra Bentz, an educator at the Feministisches Frauen Gesundheits Zentrum in Berlin, laments the way research findings on the clitoris, prostate, and ejaculation aren't passed on: "It's unbelievable that this knowledge hasn't

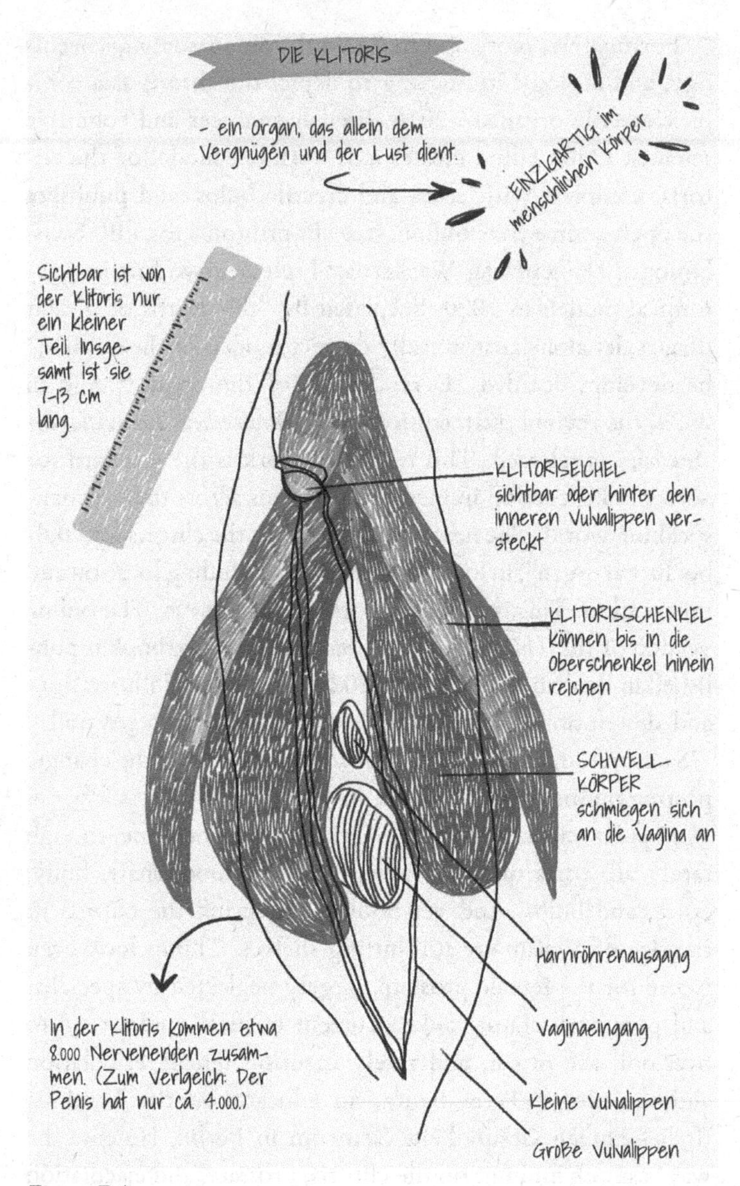

Figure E.1
Illustration from Louie Läuger's sex ed comic "*da unten*": *Über Vulven und Sexualität* (2019) (*Rethinking Gender* [2022]).

made its way into textbooks or other books or the minds of the general public. Each new generation thinks it has to reinvent the wheel. It's outrageous!"[17]

Vulval ejaculation and squirting demonstrate that dichotomies in use for more than two hundred years to describe male and female bodies along with gender characteristics are utterly absurd. Anyone who observes a squirt queen in action will readily toss out traditional notions of the female body. Active and passive, giving and receiving, strong and weak, penetrating and penetrated, the one fucking and the one being fucked—these attributes don't arise from bodies themselves but are instead cultural prescriptions in pursuit of an ideological goal. We need a new view of our bodies. We need different words. As we've seen with a term like "circlusion," language can capture newfound knowledge, shifts in perspective, and altered perception.

Figure E.2
Lucas Cranach the Elder, *The Golden Age*, ca. 1530.

Simply reversing outdated attributes, meanwhile, rarely helps. No, female coming is neither the "key to final emancipation" nor a "postfeminist duty."[18] Lots of vulva-havers don't ejaculate or soak their mattresses. The female cum shot won't be transfigured that easily into an emblem of female sexuality and self-actualization. And what good comes of putting an even greater pressure and sense of (feminist) duty on women in bed?

Sexual experiences are profoundly subjective. One critical aspect of orgasmic juice has always been the fact that it doesn't occur in all people with vulvas. The size and shape of the prostate and characteristics of the anterior vaginal wall vary significantly between individuals.[19] That's a solid explanation for why some ejaculate routinely, many only occasionally, and others never. It makes sense within the notion of diverse sexuality. Some women and vulva-havers ejaculate or/and gush, and others don't. For some it's the most fulfilling form of climax, while others find it uncomfortable, pointless, or annoying (ugh, don't tell me I have to change the sheets *again*).

How much variation will we allow our bodies? What forms of difference can we bear? When it comes to ejaculation and squirting, are we willing to accept that all are *not* equal? Even those people with vulvas who can ejaculate and gush don't always do it, meaning a purely anatomical explanation comes up short. Myriad factors influence the nature of one's orgasm, including where a person is in her cycle (if premenopausal), levels of arousal and ease, the dynamic with her partner, and the opportunity and ability to let herself go. The oldest respondent to the international online survey I've discussed, a woman of seventy-three, reports that she first ejaculated at age sixty-eight, after "she 'totally let go' during sex."[20] Relaxation is a "very important factor" for many, but the "experiential quality of a sexual encounter" may also contribute to someone's being able to ejaculate.[21]

Many women and other vulva-havers are unable to control their sexual expulsions of fluids, and who wants to lose control of their own body? Today's ideal female body is nothing if not monitored, molded, and controlled. There is no fat on this body, nor is there hair. It doesn't smell, doesn't sweat, and doesn't get wet.[22] Laurie Penny argues that "fear of female flesh and fat is fear of female power, the sublimated power of women over birth and death and dirt and sex."[23] Women are to be thin but toned, odorless, hairless, and dry.

Though 1990s' qualms about ejaculation have since abated, another no-no remains: the scent (some say stench) and squelch of female sexual juices.[24] Bodily fluids have been considered embarrassing, unpleasant, and repellent since last century. Sweat, urine, and saliva are shameful, but nothing compares to menstrual blood or female genital fluids. (Male ejaculate and sweat remain less offensive.) Potty training teaches children to hold in numbers one and two from a young age—physical control that carries into adulthood, when all manner of bodily fluids must be kept in check, neither seen nor smelled. Women are embarrassed by period leaks, pit stains, and streaks of discharge in their underwear. Many take the pill continuously to skip their period, with one reason being to avoid the worry of staining their pants.[25] Squirting and the mess it makes in bed can be a real challenge for some people.

There is a tremendous amount of pressure on women when it comes to sexuality and performance in bed. Everything we know about vulval ejaculation and squirting should effect the very opposite, though, and provide a sense of calm. Ejaculating or squirting is just one possible manifestation—among many—of female sexual power. The point is not that all women can ejaculate or gush fountains, and it's certainly not that they should. What's important is that women have the opportunity to ejaculate. That they welcome their juices. That they know what's going on when they squirt. And that they're able to recognize, name, and experience ejaculation and squirting, natural parts of female sexuality and pleasure for thousands of years. So let it flow, baby.

ACKNOWLEDGMENTS

For their many conversations, expertise, and support, the author thanks:

Petra Bentz, Feministisches Frauen Gesundheits Zentrum, Berlin

Dr. Laura Méritt, Sexclusivitäten, weiblichequelle.de, Berlin

Imbritt Wiese, Haeberle-Hirschfeld Archive of Sexology, University Library of Humboldt-Universität zu Berlin

The team at Spinnboden Lesbian Archive and Library, Berlin

And a big thank you to translator Elisabeth Lauffer, who transformed *Spritzen* into *Juice* with great skill, wit, and commitment.

IMAGE CREDITS

Figure 0.1: Deborah Sundahl, *Weibliche Ejakulation & der G-Punkt* (Emmendingen: Hans-Nietsch-Verglag, 2006). Published in English as *Female Ejaculation and the G-Spot* (Alameda, CA: Hunter House, 2003).

Figure 0.2: Peter Hesse and Günter Tembrock, eds., *Sexuologie: Geschlecht, Mensch, Gesellschaft* (Leipzig: Hirzel, 1974).

Figure 2.1: © Polly Fannlaf / Laura Méritt.

Figures 3.1 and 3.2: Thomas Laqueur, *Auf den Leib geschrieben: Die Inszenierung der Geschlechter von der Antike bis Freud* (München: Deutscher Taschenbuch-Verlag, 1996). Published in English as *Making Sex: Body and Gender from the Greeks to Freud* (Cambridge, MA: Harvard University Press, 1992).

Figure 4.1: Milan Zaviačič, *The Human Female Prostate: From Vestigal Skene's Paraurethral Glands and Ducts to Woman's Functional Prostate* (Bratislava: Slovak Academic Press, 1999).

Figure 4.2: Lynn Margulis and Dorion Sagan, *What Is Sex?* (New York: Simon & Schuster, 1997).

Figures 4.3 and 4.4: Föderation der Feministischen Frauen-Gesundheits-Zentren, eds., *Frauenkörper—neu gesehen: Ein illustriertes Handbuch* (Berlin: Orlanda Verlag, 1992). © Suzann Gage / Orlanda Verlag. Published in English as Federation of Feminist Women's Health Centers, *A New View of a*

Woman's Body: A Fully Illustrated Guide (New York: Simon and Schuster, 1981).

Figure 5.1: G-Spot: from Alice Kahn Ladas, Beverly Whipple, John D. Perry, *Der G-punkt: Das stärkste erotische Zentrum der Frau* (München: Heyne, 1983). Original publisher Holt, Rinehart & Winston. Published in English as *The G Spot and Other Recent Discoveries about Human Sexuality* (New York: Dell, 1983).

Figure 5.2: Alice Schwarzer, *Der "kleine Unterschied" und seine großen Folgen: Frauen über sich—Beginn einer Befreiung* (Frankfurt: Fischer Verlag, 1975). © Pardon.

Figure 6.1: Shannon Bell, *Whore Carnival* (New York: Autonomedia, 1995). © Annie Sprinkle.

Figure E.1: Louie Läuger, *"da unten": Über Vulven und Sexualität. Ein Aufklärungscomic* (München: Unrast Verlag, 2019). © Unrast Verlag. Published in English as *Rethinking Gender: An Illustrated Exploration* (Cambridge, MA: MIT Press, 2022).

Figure E.2: Lucas Cranach the Elder, *The Golden Age*, ca. 1530, National Museum, Oslo.

NOTES

EPIGRAPH

Marjorie Garber, *Bisexuality and the Eroticism of Everyday Life* (New York: Routledge, 2000), 75.

AUTHOR'S NOTE

1. Carmen Kloer, Augustus Parker, Gaines Blasdel, Samantha Kaplan, Lee Zhao, and Rachel Bluebond-Langner, "Sexual Health after Vaginoplasty: A Systemic Review," *Andrology* 9, no. 6 (November 2021): 1744–1764.

PREFACE

1. Cf. Florian Wimpissinger, Karl Stifter, Wolfgang Grin, and Walter Stackl, "The Female Prostate Revisited: Perineal Ultrasound and Biochemical Studies of Female Ejaculate," *Journal of Sexual Medicine* 4, no. 5 (July 2007): 1391.

2. Florian Wimpissinger, Christopher Springer, and Walter Stackl, "International Online Survey: Female Ejaculation Has a Positive Impact on Women's and Their Partners' Sexual Lives," *BJU International* 20 (2013): E177.

3. I wrote *Juice* in Germany, and though the book travels around the world in surveying more than two thousand years of cultural history, the perspective it presents is decidedly western European. Given the scope of the work, I introduce a number of topics that I am unable to present in their full complexity. I ask that the reader

bear this in mind, particularly with regard to the chapters on sexual fluids in Chinese and Indian erotic literature or the brief introduction to *kunyaza*.

4. Cf. Steven Marcus, *The Other Victorians: A Study of Sexuality and Pornography in Mid-Nineteenth-Century England* (New York: Routledge, 1964), 113.

5. Michel Foucault, *The History of Sexuality, Volume 1: An Introduction*, trans. Robert Hurley (New York: Pantheon Books, 1978), 4.

6. Laura Méritt, conversation with author, April 26, 2019.

7. Milan S. Zaviačič, Tomas Zaviačič, Richard J. Ablin, Jan Breza, and Karol Holoman, "The Female Prostate: History, Functional Morphology and Sexology Implications," *Sexologies* 11, no. 41 (2001): 44.

8. FIPAT, *Terminologia Anatomica*, 2nd ed., Federative International Programme for Anatomical Terminology, 2019, fipat.library.dal.ca /TA2/.

9. Roger Smith, *Netter's Obstetrics and Gynecology*, 2nd ed. (Philadelphia: Elsevier, 2008).

10. Female ejaculation is mentioned twice in Peter Vaupel, Hans-Georg Schaible, and Ernst Mutschler, *Anatomie, Physiologie, Pathophysiologie des Menschen* (Stuttgart: Wissenschaftliche Verlagsgesellschaft, 2015). A more detailed description can be found in Hans-Joachim Ahrendt and Cornelia Friedrich, ed., *Sexualmedizin in der Gynäkologie* (Heidelberg: Springer Verlag, 2015).

11. Vaupel, Schaible, and Mutschler, *Anatomie, Physiologie, Pathophysiologie*, 637–638.

12. Manfred Stauber and Thomas Weyerstahl, *Gynäkologie und Geburtshilfe* (Stuttgart: Thieme, 2007), 11.

13. Mary Jane Sherfey, "The Evolution and Nature of Female Sexuality in Relation to Psychoanalytic Theory," *Journal of the American Psychoanalytic Association* 14, no. 1 (January 1966): 49–50.

14. Anna Fischer-Dückelmann, *Das Geschlechtsleben des Weibes: Eine physiologisch-soziale Studie mit ärztlichen Ratschlägen* (Berlin: Hugo Bermühler Verlag, 1908), 1.

15. Quoted in Caroline Meauxsoone-Lesaffre, "L'émission fontaine ou l'éjaculation féminine," *Annales Médico-Psychologiques* 171 (2013): 110.

16. Anonymous, *Der physische Ursprung des Menschen* (Tübingen, 1800), 1:21.

17. Rebecca Chalker, *The Clitoral Truth: The Secret World at Your Fingertips* (New York: Seven Stories Press, 2000), 74.

18. Cf. Peter Hesse and Günter Tembrock, eds., *Sexuologie: Geschlecht, Mensch, Gesellschaft* (Leipzig: Hirzel, 1974), 1:211, 216. The authors also reference a "secretion reflex in the Bartholin's gland" that corresponds "approximately to the male ejaculation reflex" (216).

19. Daniel Haag-Wackernagel, "Sensorische Nervenendigungen—der Schlüssel zur weiblichen Lust," *Sexuologie* 29, nos. 1–2 (2022): 5.

20. Sylvia Groth and Kerstin Pirker, "Die Klitoris—das Lustorgan der Frau," *clio* 68 (2009): 7.

21. *Sex Education*, season 3, episode 7, directed by Kate Herron, created by Laurie Nunn, written by Sophie Goodhart, aired 2021, on Netflix, https://www.netflix.com/search?q=sex%20education&jbv=80197526.

22. Florian Wimpissinger, "Die weibliche Prostata—Faktum oder Mythos," *Urologie* 2, no. 7 (2007): 18.

23. Naomi Wolf, *Vagina: A New Biography* (New York: Ecco, 2012), 111.

24. Michael Schünke, Erik Schulte, Udo Schumacher, Markus Voll, and Karl Wesker, *Prometheus: LernAtlas der Anatomie, Innere Organe* (Stuttgart: Thieme, 2012), 311. Emphasis in original.

25. I will use this term repeatedly to remind readers that when discussing the clitoris, I am indeed referring to the organ in all of its parts.

26. Josephine Lowndes Sevely, *Eve's Secrets: A New Theory of Female Sexuality* (New York: Random House, 1987), 179.

27. Sabine zur Nieden, *Weibliche Ejakulation: Variationen zu einem uralten Streit der Geschlechter* (Gießen: Psychosozial-Verlag, 2009), 57.

28. "Now that more people are talking about it, I get the feeling more women are ejaculating," observes Austrian choreographer Christine Gaigg, whose piece on female ejaculation, titled "Go for It Let Go," ran at Tanzquartier Wien in January 2022. See Christine Gaigg and Stephanie Haerdle, "Ejaculating as a Lifestyle Thing," interview by Karin Cerny, *TQW Magazin*, last modified February 10, 2021, https://tqw.at/en/ejaculating-as-a-lifestyle-thing/.

29. Cf. Wimpissinger, Springer, and Stackl, "International Online Survey," E181. Some women report that they ejaculate from the vagina, not the urethra. This is possible given that some prostatic ducts lead to the anterior vaginal wall.

30. Cf. Wimpissinger et al., "The Female Prostate Revisited," 1391.

31. Alex Todorov, "Weibliche Ejakulation ist . . . ," JOYclub, last modified June 4, 2019, https://www.joyclub.de/magazin/sex/weibliche_ejakulation_ist.html#-squirting_die_joyclub_umfrage.

32. Alberto Rubio-Casillas and Emmanuelle A. Jannini, "New Insights from One Case of Female Ejaculation," *Journal of Sexual Medicine* 8, no. 12 (December 2011): 3500–3504; S. Salama, F. Boitrelle, A. Gauquelin, C. Lesaffre, N. Thiounn, and P. Desvaux, "Que sait-on des femmes fontaines et de l'éjaculation féminine en 2015?," *Gynécologie Obstétrique & Fertilité* 43, no. 6 (June 2015): 449–452; Zlatko Pastor and Roman Chmel, "Differential Diagnostics of Female 'Sexual' Fluids: A Narrative Review," *International Urogynecological Journal* 29, no. 4 (December 2017): 621–629. The latter study also examines coital incontinence as a third sexual fluid. See also Zlatko Pastor and Roman Chmel, "Female Ejaculation and Squirting as Similar but Completely Different Phenomena: A Narrative Review of Current Research," *Clinical Anatomy* 35, no. 5 (April 2022): 616–625.

33. Cf. Wimpissinger, Springer, and Stackl, "International Online Survey," E180.

34. Wimpissinger, Springer, and Stackl, "International Online Survey," E180.

35. Cf. Frédérique Gruyer, *Ce paradis trop violent: autour des femmes-fontaines* (Paris: R. Laffont, 1984).

36. "Sex Forum on Squirting or Female Ejaculation," JOYclub, accessed August 3, 2022, https://www.joyclub.de/forum/t483572-75.squirting_oder_die_weibliche_ejakulation.html.

CHAPTER 1

1. Cf. Robert Hans van Gulik, *Sexual Life in Ancient China: A Preliminary Survey of Chinese Sex and Society from ca. 1500 B.C. till 1644 A.D.* (Leiden: Brill, 1961), 38–39.

2. In some Taoist sects, this idea gives rise to a kind of sexual vampirism, in which a woman's (or ideally, a virginal girl's) fluids are extracted from her—a degrading practice that casts the woman as a vessel and can even lead to her death.

3. See Rudolf Pfister, "Gendering Sexual Pleasures in Early and Medieval China," *Asian Medicine* 7 (2012): 34–64.

4. Li Ling and Keith McMahon, "The Contents and Terminology of the Mawangdui Texts on the Arts of the Bedchamber," *Early China* 17 (1992): 159.

5. Rudolf Pfister, *Der beste weg unter dem himmel: Sexuelle körpertechniken aus dem alten china. Zwei bambustexte aus mawangdui* (Zürich: Museum Rietberg, 2003), 12.

6. Douglas Wile, *Art of the Bedchamber: The Chinese Sexual Yoga Classics Including Women's Solo Meditation Texts* (Albany: SUNY Press, 1992), 48; Nik Douglas and Penny Slinger, *The Erotic Sentiment in the Paintings of China and Japan* (Rochester: Park Street Press, 1990), 8.

7. Wile, *Art of the Bedchamber*, 7.

8. Pfister, *Der beste weg unter dem himmel*, 75.

9. Yimen, *Dreams of Spring: Erotic Art in China from the Bertholet Collection* (Amsterdam: Pepin Press, 1998), 23.

10. Thomas Zimmer, "Der chinesische Roman der ausgehenden Kaiserzeit," in *Geschichte der chinesischen Literatur*, ed. Wolfgang Kubin (Munich: K. G. Saur, 2002), 2:411.

11. Rudolf Pfister, *Sexuelle körpertechniken im alten China: Seimbedürftige männer im umgang mit lebens-spenderinnen: Drei manuskripte aus Mawangdui: Eine lektüre* (Norderstedt: Books on Demand, 2006).

12. Wile, *Art of the Bedchamber*, 9.

13. This contest is known as *caibu*. Cf. Zimmer, "Der chinesische Roman," 411.

14. Yimen, *Dreams of Spring*, 25.

15. Pfister, *Der beste weg unter dem himmel*, 9.

16. Pfister, *Der beste weg unter dem himmel*, 32.

17. The authorship is uncertain, and Pfister does not rule out the possibility that women were among the writers.

18. Ling and McMahon, "Mawangdui Texts," 176.

19. Pfister, *Der beste weg unter dem himmel*, 63.

20. Rudolf Pfister, "On the Partynomy of Female Genitals in Chinese Manuscripts on Sexual Body Techniques" (paper presented at the Shanghai International Conference for Medical Manuscripts Unearthed in China, May 2016), 4.

21. Ling and McMahon, "Mawangdui Texts," 178.

22. Pfister, *Der beste weg unter dem himmel*, 65, 86.

23. Rudolf Pfister, "Der Milchbaum und die Physiologie der weiblichen Ejakulation: Bemerkungen über Papiermaulbeer- und Feigenbäume im Süden Chinas," *Asiastische Studien: Zeitschrift der Schweizerischen Asiengesellschaft* 61, no. 3 (January 2007): 833.

24. Pfister, "Der Milchbaum und die Physiologie," 834, 835.

25. Robert Hans van Gulik, *Erotic Colour Prints of the Ming Period with an Essay on Chinese Sex Life from the Han to the Ch'ing Dynasty, B.C. 206–A.D. 1644* (Leiden: Brill, 2004), 44–45, 118.

26. *The Plum in the Golden Base, or Chin P'ing Mei*, trans. David Tod Roy (Princeton, NJ: Princeton University Press, 2001), 2:144–145.

27. Quoted in Zimmer, "Der chinesische Roman," 450.

28. Quoted in Fang Fu Ruan, *Sex in China: Studies in Sexology in Chinese Culture* (New York: Plenum Press, 1991), 139.

CHAPTER 2

1. Miranda Shaw, *Passionate Enlightenment: Women in Tantric Buddhism* (Princeton, NJ: Princeton University Press, 1994), 157.

2. Quoted in Renate Syed, "Zur Kenntnis der 'Gräfenberg-Zone' und der weiblichen Ejakulation in der altindischen Sexualwissenschaft: Ein medizinhistorischer Beitrag," *Sudhoffs Archiv für Geschichte der Medizin und der Naturwissenschaften* (Stuttgart: Franz Steiner Verlag, 1999): 82:181, bk. 2.

3. *Moksha*—release or liberation from the cycle of life, death, and rebirth known as *samsara*—is later added as a fourth aim.

4. Katharina Kakar and Sudhir Kakar, *The Indians: Portrait of a People* (London: Penguin Books, 2007), 73, 75. It bears mentioning that the *Kamasutra* and similar works, aimed at a largely male elite, represent a male perspective on sex.

5. *Metzler Lexikon Weltliteratur*, ed. Axel Ruckaberle (Berlin: Springer Verlag, 2006), 2:237.

6. Yashodhara Indrapada, commentary on Vatsyayana Mallanaga, *Kamasutra*, trans. Wendy Doniger and Sudhir Kakar (Oxford: Oxford University Press, 2003), 4.

7. Philip Rawson, *India* (London: Weidenfeld and Nicholson, 1968), 83–84.

8. Richard Schmidt, *Beiträge zur indischen Erotik: Das Liebesleben des Sanskritvolkes. Nach den Quellen dargestellt* (Berlin: Barsdorf, 1922), 263.

9. Cf. Joanna B. Korda, "The History of Female Ejaculation," *Journal of Sexual Medicine* 7, no. 5 (May 2010): 1967–1968.

10. Syed, "Zur Kenntnis der 'Gräfenberg-Zone,'" 183.

11. Indrapada, commentary, 35.

12. Kokkoka, *Rati Rahasya of Pandit Kokkoka*, trans. S. C. Upadhyaya (Bombay: D. B. Taraporevala, 1965), 29.

13. Kokkoka, *Rati Rahasya of Pandit Kokkoka*, 21–23, 45.

14. Korda, "The History of Female Ejaculation," 1968.

15. Kalayana Malla, *Ananga Ranga: Stage of the Bodiless One, or The Hindu Art of Love*, trans. F. F. Arbuthnot and Richard F. Burton (New York: Medical Press of New York, 1964), 4, 5.

16. Malla, *Ananga Ranga*, 125, 38. "It is for instance, clearly evident that unless by some act of artifice the venereal orgasm of the female, who is colder in blood and less easily excited, distinctly precede that of the male, the congress has been in vain, the labour of the latter has done no good, and the former has enjoyed no satisfaction. Hence it results that one of man's chief duties in this life is to learn to withhold himself as much as possible, and, at the same time, to hasten the enjoyment of his partner."

17. Schmidt, *Beiträge zur indischen Erotik*, 645.

18. Kokkoka, *Rati rahasya of Pandit Kokkoka*, 54.

19. Korda, "The History of Female Ejaculation," 1969.

20. Kokkoka, *Rati rahasya of Pandit Kokkoka*, 54.

21. Schmidt, *Beiträge zur indischen Erotik*, 258–259.

22. Syed, "Zur Kenntnis der 'Gräfenberg-Zone,'" 178, 171.

23. Mark S. G. Dyczkowski, *The Cult of the Goddess Kubjika: A Preliminary Comparative Textual and Anthropological Survey of a Secret Newar Goddess* (Stuttgart: Franz Steiner Verlag, 2001), 39.

24. Cf. David Gordon White, *Kiss of the Yoginī: "Tantric Sex" in Its South Asian Contexts* (Chicago: University of Chicago Press, 2006), 93.

25. Dyczkowski, *The Cult of the Goddess Kubjika*, 47.

26. Cf. White, *Kiss of the Yoginī*, 93.

27. Rufus Camphausen, *The Yoni: Sacred Symbol of Female Creative Power* (Rochester, VT: Inner Traditions, 1996). Camphausen's book provides an introduction to Indian Tantric rituals, still practiced today, in which vaginal fluids are collected and drunk.

28. Quoted in Miranda Shaw, *Passionate Enlightenment: Women in Tantric Buddhism* (Princeton, NJ: Princeton University Press, 1994), 39.

29. Shaw, *Passionate Enlightenment*, 168.

30. Shaw, *Passionate Enlightenment*, 157.

31. Quoted in Shaw, *Passionate Enlightenment*, 158.

32. Shaw, *Passionate Enlightenment*, 158. Cf. Catherine Blackledge, *The Story of V: Opening Pandora's Box* (London: Weidenfeld and Nicholson, 2003), 265–266.

33. Cf. Camphausen, *The Yoni*.

34. Shaw, *Passionate Enlightenment*, 158.

35. Cf. White, *Kiss of the Yoginī*, 75.

36. White, *Kiss of the Yoginī*, 75.

37. Cf. Kakar and Kakar, *The Indians*. This work outlines various interpretations of male semen in ancient India.

38. Cf. Schmidt, *Beiträge zur indischen Erotik*, 285–286.

39. Cf. Jeffrey Gettleman, "The Peculiar Position of India's Third Gender," *New York Times*, February 17, 2018, https://www.nytimes.com/2018/02/17/style/india-third-gender-hijras-transgender.html.

40. Cf. Schmidt, *Beiträge zur indischen Erotik*, 286.

41. Cf. White, *Kiss of the Yoginī*, 92.

CHAPTER 3

1. Cf. Karl Stifter, *Die dritte Dimension der Lust: Das Geheimnis der weiblichen Ejakulation* (Munich: Heyne Verlag, 1988), 30.

2. Thomas Laqueur, *Making Sex: Body and Gender from the Greeks to Freud* (Cambridge, MA: Harvard University Press, 1992), 52.

3. Laqueur, *Making Sex*, 4–5. Claudius Galenus (129–199 CE), whose medical authority holds into the early modern period, refers to the ovaries and testicles by the same term, *orcheis*.

4. Laqueur, *Making Sex*, 38.

5. Heinz-Jürgen Voß, *Making Sex Revisited. Dekonstruktion des Geschlechts aus biologisch-medizinischer Perspektive* (Bielefeld: Transcript,

2010), 52–60. The author thoroughly outlines ancient Greek and Roman teachings on semen.

6. Cf. Stifter, *Die dritte Dimension der Lust*, 28.

7. Hippocrates of Kos, *Generation*, Loeb Classical Library 520:12–13, accessed July 21, 2023, https://www.loebclassics.com/view/hippocrates_cos-generation/2012/pb_LCL520.13.xml?result=27&rskey=F3DP3R. As far as I know, this concept was never adopted or developed further by other philosophers or physicians.

8. Hippocrates of Kos, *Generation*.

9. Stifter, *Die dritte Dimension der Lust*, 33.

10. Quoted in Giulia Sissa, "The Sexual Philosophies of Plato and Aristotle," in *From Ancient Goddesses to Christian Saints*, trans. Arthur Goldhammer, ed. Pauline Schmitt Pantel, vol. 1, *A History of Women in the West*, ed. Georges Duby and Michelle Perrot (Cambridge, MA: Belknap Press of Harvard University Press, 1992), 68–69.

11. Aristotle, *Generation of Animals*, Loeb Classical Library 366:100–101, accessed July 21, 2023, https://www.loebclassics.com/view/aristotle-generation_animals/1942/pb_LCL366.101.xml?rskey=V5xOxM&result=2&mainRsKey=JztsrH.

12. Laqueur, *Making Sex*, 4.

13. Galen, *On the Usefulness of the Parts of the Body*, trans. Margaret Talladge May (Ithaca, NY: Cornell University Press, 1968), 14.II.297.

14. Galen, *On the Usefulness of the Parts of the Body*.

15. Thomas Laqueur, *Solitary Sex: A Cultural History of Masturbation* (New York: Zone Books, 2003), 93. Brackets in original.

16. Norbert Angermann, ed., *Lexikon des Mittelalters* (Munich: Artemis Verlag, 1995), 7:1817.

17. Robert Jütte, *Lust ohne Last: Geschichte der Empfängnisverhütung* (Munich: C. H. Beck Verlag, 2003), 110.

18. Julius Jörimann, *Frühmittelalterliche Rezeptarien* (Zürich, 1925), 153.

19. Britta-Juliane Kruse, *Verborgene Heilkünste: Geschichte der Frauenmedizin im Spätmittelalter* (Berlin: De Gruyter, 1996), 229. "Vnd wen der frawen samet herschet ūber des mannes samen so wirt es ein meitlin hers[ch]et aber des mannes samen über der frawen so wirt es ein knab."

20. Paul Diepgen, *Frau und Frauenheilkunde in der Kultur des Mittelalters* (Stuttgart: Thieme, 1963), 74.

21. Kruse, *Verborgene Heilkünste*, 240.

22. Wolfgang Gerlach, "Das Problem des 'weiblichen Samens' in der antiken und mittelalterlichen Medizin," *Sudhoffs Archiv für Geschichte der Medizin und der Naturwissenschaften* (February 1939): 30:191–193, bks. 4–5.

23. Quoted in Sabine zur Nieden, *Weibliche Ejakulation: Variationen zu einem uralten Streit der Geschlechter* (Gießen: Psychosozial-Verlag, 2009), 41.

24. Gerlach, "Das Problem des 'weiblichen Samens,'" 192.

25. Cf. Jean-Louis Flandrin, "Das Geschlechtsleben der Eheleute in der alten Gesellschaft: Von der kirchlichen Lehre zum realen Verhalten," in *Die Masken des Begehrens und die Metamorphosen der Sinnlichkeit: Zur Geschichte der Sexualität im Abendland*, ed. Philippe Ariès and André Béjin (Frankfurt: Fischer Wissenschaft, 1984), 151–152.

26. Steven Ozment, *When Fathers Ruled: Family Life in Reformation Europe* (Cambridge, MA: Harvard University Press, 1983), 216.

27. Wolfgang Ertler, *Im Rausch der Sinnlichkeit: Die Geschichte der unterdrückten Lust und die Vision einer paradiesischen Sexualität* (Munich: Diederichs, 2001), 120.

28. Josephine Lowndes Sevely, *Eve's Secrets: A New Theory of Female Sexuality* (New York: Random House, 1987), 58.

29. Cf. Pieter Willem van der Horst, "Sarah's Seminal Emissions: Hebrews 11:11 in the Light of Ancient Embryology," in *Hellenism-Judaism-Christianity: Essays on Their Interaction* (Leuven: Peeters, 1998). Van der Horst has found that the Greek concept of female semen was familiar to early Jews as well.

30. Cf. Claude Thomasset, "The Nature of Woman," in *Silences of the Middle Ages*, trans. Arthur Goldhammer, ed. Christiane Klapisch-Zuber, vol. 2, *A History of Women in the West*, ed. Georges Duby and Michelle Perrot (Cambridge, MA: Belknap Press of Harvard University Press, 1992), 58; Angermann, *Lexikon des Mittelalters*, 1817.

31. Quoted in Mark D. Stringer, "Colombo and the Clitoris," *European Journal of Obstetrics* 151, no. 2 (August 2010): 131.

32. Thomasset, "The Nature of Woman," 56.

33. Cf. Angermann, *Lexikon des Mittelalters*, 1817.

34. Kruse, *Verborgene Heilkünste*, 228.

35. Cf. Chris E. Paschold, *Die Frau und ihr Körper im medizinischen und didaktischen Schrifttum des französischen Mittelalters: Wortgeschichtliche Untersuchungen zu Texten des 13. und 14. Jahrhunderts. Mit kritischer Ausgabe der gynäkologischen Kapitel aus den "Amphorismes Ypocras" des Martin de Saint-Gilles* (Hannover, 1986), 89.

36. Cf. Angermann, *Lexikon des Mittelalters*, 1819.

37. Abū ʿAbdallāh Muḥammad an-Nafzāwī, *The Perfumed Garden of the Cheikh Nefzaoui: A Manual of Arabian Erotology*, trans. Richard Francis Burton (New York: Signet Classic, 1999), 68.

38. Quoted in Konrad Goehl, *Frauengeheimnisse im Mittelalter: Die Frauen von Salern* (Baden-Baden: Deutscher Wissenschafts-Verlag, 2010), 8.

39. Stifter, *Die dritte Dimension der Lust*, 50, 51.

40. Angermann, *Lexikon des Mittelalters*, 1818.

41. Kruse, *Verborgene Heilkünste*, 93–94, 105–106.

42. Diepgen, *Frau und Frauenheilkunde*, 181.

43. Goehl, *Frauengeheimnisse im Mittelalter*, 19.

44. Quoted in Évelyne Berriot-Salvadore, "The Discourse of Medicine and Science," in *Renaissance and Enlightenment Paradoxes*, trans. Arthur Goldhammer, ed. Natalie Zemon Davis and Arlette Farge, vol. 3, *A History of Women in the West*, ed. Georges Duby and Michelle

Perrot (Cambridge, MA: Belknap Press of Harvard University Press, 1993), 362.

45. Diepgen, *Frau und Frauenheilkunde*, 183.

46. Goehl, *Frauengeheimnisse im Mittelalter*, 20.

47. Robert Muchembled, *Orgasm and the West: A History of Pleasure from the 16th Century to the Present*, trans. Jean Birrell (Cambridge, UK: Polity Press, 2008), 73.

48. Barbara Duden, *Geschichte unter der Haut: Ein Eisenacher Arzt und seine Patientinnen um 1730* (Stuttgart: Klett-Cotta, 1991), 138.

49. Diepgen, *Frau und Frauenheilkunde*, 173.

50. Stifter, *Die dritte Dimension der Lust*, 43.

51. Cf. Ursula Weisser, *Zeugung, Vererbung und pränatale Entwicklung in der Medizin des arabisch-islamischen Mittelalters* (Erlangen: Lüling, 1983), 130–131.

52. Basim F. Musallam, *Sex and Society in Islam: Birth Control before the Nineteenth Century* (Cambridge, UK: Cambridge University Press, 1983), 51.

53. Esin Kahya, ed., *The Treatise on Anatomy of Human Body and Interpretation of Philosophers by Al-'Itaqi* (Islamabad: National Hijra Council [Pakistan], 1990), 121, 117.

54. Musallam, *Sex and Society in Islam*, 17.

55. Cf. Musallam, *Sex and Society in Islam*, 61–62.

56. Musallam, *Sex and Society in Islam*, 61.

57. Cf. Musallam, *Sex and Society in Islam*, 64.

58. Quoted in Wadie B. Chenouda, "Die Abhandlung von Milz, Nieren, Harnblase, Hoden, Penis, Gebärmutter und Brust nach dem 'Handbuch der Chirurgie des Ibn al-Quff'" (PhD diss., Julius-Maximilians-Universität of Würzburg, 1988), 48.

59. An-Nafzāwī, *The Perfumed Garden*, 11, 18, 65.

60. Cf. Ali Ghandour, *Liebe, Sex und Allah: Das unterdrückte erotische Erbe der Muslime* (Munich: C. H. Beck Verlag, 2019), 137.

61. Ghandour, *Liebe, Sex und Allah*, 117, 110. Brackets in original, translated from German by Elisabeth Lauffer.

62. An-Nafzāwī, *The Perfumed Garden*, 130–131, 64–65.

63. Nicolas Venette, *The Mysteries of Conjugal Love Reveal'd* (London, 1712), 143.

64. Venette, *The Mysteries of Conjugal Love Reveal'd*, 222, 223–224, 231.

65. Venette, *The Mysteries of Conjugal Love Reveal'd*, 99, 140.

66. Jütte, *Lust ohne Last*, 108–109.

67. Samuel Auguste David Tissot, *Onanism, or, A Treatise upon the Disorders Produced by Masturbation, or, The Dangerous Effects of Secret and Excessive Venery*, trans. A. Hume (London: printed for Richardson and Urquhart, 1781), 52.

68. Tissot, *Onanism*, 46, 71.

69. Laqueur, *Solitary Sex*, 39.

70. Tissot, *Onanism*, 161.

71. Simon André Tissot, *Von der Onanie oder Abhandlung ueber die Krankheiten, die von der Selbstbefleckung herrühren* (Eisenach: Michael Gottlieb Grießbach's sel. Söhnen, 1770), 221–222.

72. Thomas Laqueur, "Teufelszeug," interview by Martin Spiewak, *Die Zeit*, no. 17, April 17, 2008.

73. Johann Heinrich Zedler, *Grosses vollstaendiges Universal-Lexicon aller Wissenschafften und Künste* (Halle: J. H. Zedler, 1739), 10:639. Emphasis in original.

74. Zedler, *Grosses vollstaendiges Universal-Lexicon*, 22:1652. Emphasis in original.

75. Laqueur, *Making Sex*, 149–150.

76. Dietrich Wilhelm Heinrich Busch, *Das Geschlechtsleben des Weibes in physiologischer, pathologischer und therapeutischer Sicht* (Leipzig: Brockhaus, 1839), 60.

77. Duden, *Geschichte unter der Haut*, 20–21, 34–35.

78. Dr. Ploss-Barthels, quoted in Emil Seyler, *Die Frau des XX: Jahrhunderts und ihre Krankheiten* (Leipzig: Verlag Otto Borggold, 1900), 39.

79. Quoted in Claudia Honegger, *Die Ordnung der Geschlechter: Die Wissenschaften vom Menschen und das Weib 1750–1850* (Frankfurt: Campus Verlag, 1996), 206.

80. Johann Christian Gottfried Jörg and Heinrich Gottlieb Tzschirner, *Die Ehe aus dem Gesichtspunkte der Natur, der Moral und der Kirche betrachtet* (Leipzig: Baumgärtner Buchhandlung, 1918), 23, 58.

81. Helene von Druskowitz, *Der Mann als logische und sittliche Unmöglichkeit und als Fluch der Welt: Pessimistische Kardinalsätze* (Freiburg: Kore Verlag, 1988), 80, 34.

82. Quoted in Marita Metz-Becker, "Akademische Geburtshilfe im 19. Jahrhundert: Der Blick des Arztes auf die Frau," in *Hebammenkunst gestern und heute: Zur Kultur des Gebärens durch drei Jahrhunderte*, ed. Anika Bierig and Marita Metz-Becker (Marburg: Jonas Verlag, 1999), 40–41.

83. Laqueur, *Making Sex*, 149.

84. Honegger, *Die Ordnung der Geschlechter*, 182.

85. Busch, *Das Geschlechtsleben des Weibes*, 187.

86. Laqueur, *Making Sex*.

87. Quoted in Stifter, *Die dritte Dimension der Lust*, 63.

88. Cf. Josephine Lowndes Sevely and J. W. Bennett, "Concerning Female Ejaculation and the Female Prostate," *Journal of Sex Research* 14, no. 1 (February 1978): 1–20; Theodor Billroth and Georg Albert Luecke, *Handbuch der Frauenkrankheiten* (Stuttgart: Ferdinand Enke Verlag, 1885).

89. Cf. Laqueur, *Making Sex*.

90. Heinrich von Kleist, *The Marquise of O*, trans. Nicholas Jacobs (London: Pushkin Press, 2020).

91. Paolo Mantegazza, *Die Hygiene der Liebe*, trans. R. Teuscher (Berlin: Neufeld und Henius, 1877), 193.

92. Laqueur, *Making Sex*, 3.

93. Quoted in Laqueur, *Making Sex*, 190.

94. Christa Putz, *Verordnete Lust: Sexualmedizin, Psychoanalyse und die "Krise der Ehe," 1870–1930* (Bielefeld: Transcript, 2011), 48.

95. Muchembled, *Orgasm and the West*, 96.

96. Carolin S. Fischer, *Gärten der Lust: Eine Geschichte erregender Lektüren* (Munich: dtv, 2000), 182.

97. "Sunday, 9 February 1667/68," The Diary of Samuel Pepys, accessed May 2, 2018, https://www.pepysdiary.com/diary/1668/02/09/.

98. Laqueur, *Solitary Sex*, 334; Fischer, *Gärten der Lust*, 183.

99. Muchembled, *Orgasm and the West*, 99.

100. Anonymous, *The School of Venus, or The Ladies Delight, Reduced into Rules of Practice* (1860), 14.

101. Muchembled, *Orgasm and the West*, 96.

102. *The School of Venus*, 20–21.

103. Nicolas Chorier, *A Dialogue between a Married Lady and a Maid* (London, 1710), 42.

104. Eberhard Kronhausen and Phyllis Kronhausen, *Bücher aus dem Giftschrank: Eine Analyse der verbotenen und verfemten erotischen Literatur* (Bern: Rütten + Loening, 1969), 247.

105. Wayland Young, *Eros Denied: Sex in Western Society* (New York: Grove Press, 1964), 325–326. Quoted in Sevely and Bennett, "Concerning Female Ejaculation," 7.

106. Anonymous, *My Secret Life* (Amsterdam: privately printed for subscribers, 1888), 1:127.

107. Chorier, *A Dialogue*, 36.

108. Cleland, *Memoirs of Fanny Hill*, 320; *My Secret Life*; Anonymous, *Josefine Mutzenbacher or, The Story of a Viennese Whore as Told by Herself* (Vienna, 1906).

CHAPTER 4

1. Outside the medical discourse, "pollutions" are associated with men. Definitive reference works of the time, like *Webster's Complete Dictionary of the English Language* (1886), define pollutions as the "emission of semen, or sperm, at other times than in sexual intercourse." *Meyers Großes Konversations-Lexikon* (1907) informs German speakers that the term means the "involuntary loss of semen" in "sexually mature, abstinent men."

2. Volkmar Sigusch, *Geschichte der Sexualwissenschaft* (Frankfurt: Campus Verlag, 2008), 53.

3. Paolo Mantegazza, "Die Physiologie des Genusses," in *Gesammelte Schriften* (Berlin, 1893), 20.

4. Paolo Mantegazza, *Die Hygiene der Liebe*, trans. R. Teuscher (Berlin: Neufeld und Henius, 1877), 186.

5. Moriz Rosenthal, *A Clinical Treatise on the Diseases of the Nervous System*, trans. Leopold Putzel (London: Sampson Low, Marston, Searle, and Rivington, 1881), 2:44.

6. Cf. Sheila Jeffreys, *The Spinster and Her Enemies: Feminism and Sexuality 1880–1930* (London: Pandora Press, 1985), 110. Twentieth-century British feminist Sheila Jeffreys rejects female ejaculation as a male invention and sexual fantasy about lesbians. She names Krafft-Ebing as a prime perpetrator.

7. Richard von Krafft-Ebing, "Ueber pollutionsartige Vorgänge beim Weibe," *Neue Medizinische Press* 14 (1888); *Zeitschrift für Sexualforschung* 1 (March 1991): 70–71.

8. Krafft-Ebing, "Ueber pollutionsartige Vorgänge," 72.

9. Krafft-Ebing, "Ueber pollutionsartige Vorgänge," 33, 12–13.

10. Krafft-Ebing, "Ueber pollutionsartige Vorgänge," 33.

11. Richard von Krafft-Ebing, *Psychopathia Sexualis*, trans. Charles Gilbert Chaddock (Philadelphia: F. A. Davis Company, 1894), 31.

12. Volkmar Sigusch and Günter Grau, eds., *Personenlexikon der Sexualforschung* (Frankfurt: Campus Verlag, 2009), 154.

13. Albert Eulenburg, *Sexuale Neuropathie: Genitale Neurosen und Neuropsychosen der Männer und Frauen* (Leipzig: Verlag von F. C. Vogel, 1895), 75.

14. Eulenburg, *Sexuale Neuropathie*, 32, 82.

15. Christa Putz, *Verordnete Lust: Sexualmedizin, Psychoanalyse und die "Krise der Ehe," 1870–1930* (Bielefeld: Transcript, 2011), 187.

16. Including Krafft-Ebing, "Ueber pollutionsartige Vorgänge." Cf. Putz, *Verordnete Lust*, 115–116.

17. Quoted in Sabine zur Nieden, *Weibliche Ejakulation: Variationen zu einem uralten Streit der Geschlechter* (Gießen: Psychosozial-Verlag, 2009), 45.

18. Enoch Heinrich Kisch, *The Sexual Life of Woman in Its Physiological, Pathological and Hygienic Aspects*, trans. M. Eden Paul (New York: Rebman Company, 1910), 357.

19. Kisch, *The Sexual Life of Woman*, 357, 351. L. H. von Guttceit is among the first to use the term, in 1870, to denote female orgasm. In 1923, Albert Moll documents the instance of women who experience the sensation of ejaculation without heightened desire. Cf. Putz, *Verordnete Lust*, 48, 81.

20. In 1903, doctor Paul Bernhardt also describes two cases of female pollutions that occurred without the feeling of pleasure. Paul Bernhardt, "Ueber pollutionsartige Vorgänge beim Weibe ohne sexuelle Vorstellungen und Lustgefühle," *Neurologisches Centralblatt* 4, ed. Toby Cohn (February 2, 1905): 171.

21. Kisch, *The Sexual Life of Woman*, 357, 359, 358.

22. Cf. Putz, *Verordnete Lust*, 49.

23. Otto Adler, *Die mangelhafte Geschlechtsempfindung des Weibes: Anaesthesia sexualis feminarum. Anaphrodisia, Dyspareunia* (Berlin: Fischers medicinische Buchhandlung, 1919), 127.

24. Adler, *Die mangelhafte Geschlechtsempfindung des Weibes*, 6, 10, 16, 129.

25. Adler, *Die mangelhafte Geschlechtsempfindung des Weibes*, 16, 19.

26. Adler, *Die mangelhafte Geschlechtsempfindung des Weibes*, 23, 107, 45.

27. Adler, *Die mangelhafte Geschlechtsempfindung des Weibes*, 20.

28. Cf. Wilhelm Stekel, *Onanie und Homosexualität: Die homosexuelle Neurose* (Berlin: Urban und Schwarzenberg, 1917), 41.

29. Cf. Sigusch, *Geschichte der Sexualwissenschaft*, 208.

30. Sigusch, *Geschichte der Sexualwissenschaft*, 58.

31. Quoted in Karl Stifter, *Die dritte Dimension der Lust: Das Geheimnis der weiblichen Ejakulation* (Munich: Heyne Verlag, 1988), 107.

32. Alfred C. Kinsey, *Sexual Behavior in the Human Female* (Philadelphia: W. B. Saunders Company, 1953), 635, 634.

33. Kisch, *The Sexual Life of Woman*, 357.

34. Enoch Heinrich Kisch, *Die sexuelle Untreue der Frau* (Bonn: A. Marcus und E. Webers Verlag, 1918), 80, 81–82.

35. Kisch, *Die sexuelle Untreue der Frau*, 82.

36. Thomas Laqueur, *Making Sex: Body and Gender from the Greeks to Freud* (Cambridge, MA: Harvard University Press, 1992), viii.

37. Quoted in Putz, *Verordnete Lust*, 90.

38. Magnus Hirschfeld, *The Homosexuality of Men and Women*, trans. Michael A. Lombardi-Nash (Amherst, NY: Prometheus Books, 2000).

39. Bernhard A. Bauer, *Wie bist du, Weib? Betrachtungen über Körper, Seele, Sexualleben und Erotik des Weibes* (Vienna: Rikola, 1923), 236. Translated by Elisabeth Lauffer. Cf. Stifter, *Die dritte Dimension der Lust*, 111–114. Stifter also remarks on the "fairy tale of catapulting cervical plugs."

40. Bauer, *Wie bist du, Weib?*, 236–237.

41. Putz, *Verordnete Lust*, 233.

42. Wilhelm Stekel, *Frigidity in Woman in Relation to Her Love Life*, trans. James S. van Teslaar (New York: Liveright, 1926), 2:302. Emphasis in original.

43. Georg Ludwig Kobelt, *Die männlichen und weiblichen Wollust-Organe des Menschen und einiger Säugethiere in anatomisch-physiologischer Beziehung* (Freiburg: Emmerling, 1844), 13. He writes, "For the same reason, it is difficult for the man to make water with an erection, whereas involuntary micturition during estrus is not uncommon in the woman."

44. Quoted in Stifter, *Die dritte Dimension der Lust*, 115.

45. Ernst Gräfenberg, "The Role of Urethra in Female Orgasm," *International Journal of Sexology* 3, no. 3 (February 1950): 147.

46. William H. Masters, Virginia E. Johnson, and Robert C. Kolodny, *Masters and Johnson on Sex and Human Loving* (Boston: Little, Brown and Company, 1985), 70.

47. Cf. Beverly Whipple, "Ejaculation, Female," in *The International Encyclopedia of Human Sexuality*, ed. Patricia Whelehan and Anne Bolin (Chichester, UK: Wiley-Blackwell, 2015), 1:325.

48. Cf. Nieden, *Weibliche Ejakulation*, 46–47.

49. Quoted in Shannon Bell, "Feminist Ejaculations," in *The Hysterical Male: New Feminist Theory*, ed. Arthur Kroker and Marielouise Kroker (New York: St. Martin's Press, 1991), 159.

50. Joseph G. Bohlen, "'Female Ejaculation' and Urinary Stress Incontinence," *Journal of Sexual Research* 18, no. 4 (November 1982): 360.

51. Eve Ensler, *The Vagina Monologues* (New York: Villard, 1998), 27.

52. Rebecca Chalker, *The Clitoral Truth: The Secret World at Your Fingertips* (New York: Seven Stories Press, 2000).

53. In a systematic literature review in 2013, Zlatko Pastor concludes that wet orgasms can be explained in one of two ways: women are either ejaculating a small amount of prostatic fluid or squirting "a larger amount of diluted and changed urine." See Zlatko Pastor, "Female Ejaculation Orgasm vs. Coital Incontinence: A Systematic Review," *Journal of Sexual Medicine* 10, no. 7 (July 2013): 1682. In 2014, Samuel Salama and colleagues publish a study of seven women who report expressing large amounts of fluid during sex. The

researchers examine the fluid and conclude "that squirting is essentially the involuntary emission of urine during sexual activity." Of the seven postcoital urine samples collected, it's worth noting that five contained PSA. See Samuel Salama, Florence Boitrelle, Amélie Gauquelin, Lydia Malagrida, Nicolas Thiounn, and Pierre Desvaux, "Nature and Origin of 'Squirting' in Female Sexuality," *Journal of Sexual Medicine* 12, no. 3 (March 2015): 661. And in 2016, Ihab Younis and Rehab M. Salem assert that women who come during sex can be grouped into three categories: ejaculators, squirting women, and women with coital urinary incontinence. See Ihab Younis and Rehab M. Salem, "Female Ejaculation: Who Is Going to Sleep on the Wet Side of the Bed?," *Human Andrology* 6, no. 3 (September 2016): 90.

54. Rufus Cartwright, Susannah Elvy, and Linda Cardozo, "Do Women with Female Ejaculation Have Detrusor Overactivity?," *Journal of Sexual Medicine* 4, no. 6 (November 2007): 1655.

55. Jessica Påfs, "A Sexual Superpower or a Shame? Women's Diverging Experiences of Squirting/Female Ejaculation in Sweden," *Sexualities* (September 2021): 10.

56. Cf. B. A. Simon and T. Rokyo, "The Female Prostate," *Anthropologie* 33, nos. 1–2 (1995): 131–134.

57. Cf. Galen, *On Semen*, trans. and ed. Phillip De Lacy (Berlin: Akademie Verlag, 1992), 197–207.

58. Ericka Johnson, *A Cultural Biography of the Prostate* (Cambridge, MA: MIT Press, 2021), 25.

59. Stifter, *Die dritte Dimension der Lust*, 65.

60. G. Oberdieck, "Ueber Epithel und Drüsen der Harnblase und der männlichen und weiblichen Urethra" (PhD diss., University of Göttingen, 1884); Ludwig Aschoff, "Ein Beitrag zur normalen und pathologischen Anatomie der Schleimhaut der Harnwege und ihrer drüsigen Anhänge," *Archiv für pathologische Anatomie und Physiologie und für klinische Medicin* 138, no. 2 (November 1894): 195–220.

61. Gustaf Pallin, "Beitrag zur Anatomie und Embryologie der Prostata und der Samenblasen," in *Archiv für Anatomie und Physio-*

logie (Leipzig: Veit und Comp., 1901), 135–176; Walther Felix, *Die Entwicklung der Harn- und Geschlechtsorgane* (Leipzig: Hirzel, 1911), n.p.; Josephine Lowndes Sevely, *Eve's Secrets: A New Theory of Female Sexuality* (New York: Random House, 1987), 76.

62. E. N. Petrowa, C. S. Karaewa, and A. E. Berkowskaja, "Über den Bau der weiblichen Urethra," *Archiv für Gynäkologie* 136, no. 1 (December 1937): 343–357.

63. Quoted in Nieden, *Weibliche Ejakulation*, 34.

64. Russell L. Deter, George T. Caldwell, and A. I. Folsom, "A Clinical and Pathological Study of the Posterior Female Urethra," *Journal of Urology* 55, no. 6 (June 1946): 651–662, 653.

65. Huffman studies the history of the prostate's anatomical representation. His 1948 article analyzes fourteen anatomical accounts from 1737 through the first half of the twentieth century, all of which confirm the existence of the female prostate.

66. Cf. Regnier de Graaf, *De mulierum organis in generationi inservientibus tractatus novus* (Lugduni Batav: Ex officinas Hackiana, 1672).

67. Quoted in Sevely, *Eve's Secrets*, 70, 71.

68. Regnier de Graaf, *New Treatise concerning the Generative Organs of Women* (1672), in H. B. Jocelyn and B. P. Setchell, trans., *Journal of Reproduction and Fertility*, supplement 17 (Oxford: Blackwell Scientific Publications, 1972), 77–222.

69. The Bartholin's glands, which resemble cannellini beans, are today considered the female homologue to the Cowper's glands (aka bulbo-urethral glands) and secrete a few drops of fluid during sex, but before orgasm. The paired ducts are half to one inch long, and empty into the vulval vestibule, between the labia minora, positioned at "8 and 4 o'clock."

70. Stifter, *Die dritte Dimension der Lust*, 60–61.

71. Friedrich Tiedemann, *Von den Duverneyschen, Bartholinschen oder Cowperschen Drüsen des Weibs und der schiefen Gestaltung und Lage der Gebärmutter* (Heidelberg: Druck und Verlag von Karl Groos, 1840), 11.

72. Stifter, *Die dritte Dimension der Lust*, 100, 101, 60.

73. Alexander Skene, *The Anatomy and Pathology of Two Important Glands of the Female Urethra* (New York: William Wood and Co., 1880), 1–2.

74. Nieden, *Weibliche Ejakulation*, 34.

75. Florian Wimpissinger, "Die weibliche Prostata—Faktum oder Mythos," *Urologie* 2, no. 7 (2007): 19.

76. John W. Huffman, "The Detailed Anatomy of the Paraurethral Ducts in the Adult Human Female," *American Journal of Obstetrics and Gynecology* 55, no. 1 (January 1948): 98.

77. Wimpissinger, "Die weibliche Prostata," 20.

78. Matthias David, Frank C. K. Chen, and Peter Siedentopf, "Wer (er)fand den G-Punkt? Medizinhistorische Anmerkungen zur Erstbeschreibung vor 61 Jahren," *Deutsches Ärzteblatt* 42 (2005): A 2854.

79. Bell, "Feminist Ejaculations," 163.

80. Gräfenberg, "The Role of Urethra in Female Orgasm," 148.

81. William H. Masters and Virginia E. Johnson, *Human Sexual Response* (Boston: Little, Brown and Company, 1966), 135.

82. Sherfey's work largely targeted Freudian thought and contemporary psychoanalysis.

83. Mary Jane Sherfey, *The Nature and Evolution of Female Sexuality* (New York: Random House, 1966), 174, 14.

84. Sherfey, *The Nature and Evolution of Female Sexuality*, 38, 47, 46, 58.

85. Sherfey, *The Nature and Evolution of Female Sexuality*, 64.

86. Sherfey, *The Nature and Evolution of Female Sexuality*, 85, 88. Emphasis in original.

87. Josephine Lowndes Sevely and J. W. Bennett, "Concerning Female Ejaculation and the Female Prostate," *Journal of Sex Research* 14, no. 1 (February 1978): 18.

88. *Hexengeflüster: Frauen greifen zur Selbsthilfe* (1975), a central publication of the West German women's movement, does not mention the female prostate or ejaculation.

89. Chalker, *The Clitoral Truth*, 34.

90. Federation of Feminist Women's Health Centers, *A New View of a Woman's Body: A Fully Illustrated Guide* (New York: Simon and Schuster, 1981), 47.

91. Quoted in Chalker, *The Clitoral Truth*, 43.

92. Quoted in Shannon Bell, *Whore Carnival* (New York: Autonomedia, 1995), 273.

93. Federation of Feminist Women's Health Centers, *A New View*, 54. Pat Califia's classic *Sapphistry: The Book of Lesbian Sexuality* was published in 1980. Califia describes a "lubrication problem I've often heard lesbians talk about"—female ejaculation or squirting. Unable to explain the phenomenon, and since women's "sudden secretions startle or upset their partners," she recommends seeking medical attention. Could it be a sign of infection or incontinence? Should women use a tampon to catch the fluid? Although Califia assures readers, "You're not alone," it's easy to imagine how much relief *A New View of a Woman's Body* provided. See Pat Califia, *Sapphistry: The Book of Lesbian Sexuality* (Tallahassee, FL: Naiad Press, 1980).

94. Feministisches Frauen Gesundheits Zentrum Berlin-West, foreword to *Frauenkörper—neu gesehen: Ein illustriertes Handbuch*, ed. Föderation der Feministischen Frauen-Gesundheits-Zentren (Berlin: Orlanda-Frauenverlag, 1992), n.p.

95. Federation of Feminist Women's Health Centers, *A New View*, 54.

CHAPTER 5

1. Alice Kahn Ladas, Beverly Whipple, and John D. Perry, *The G Spot and Other Recent Discoveries about Human Sexuality* (New York: Dell, 1983), xvi, xv.

2. Janice M. Irvine, *Disorders of Desire: Sexuality and Gender in Modern American Sexology* (Philadelphia: Temple University Press, 2005), 169.

3. Ladas, Whipple, and Perry, *The G Spot*, 170.

4. John Heidenry, *What Wild Ecstasy: The Rise and Fall of the Sexual Revolution* (New York: Simon and Schuster, 1997), 301.

5. Irvine, *Disorders of Desire*, 168.

6. Ladas, Whipple, and Perry, *The G Spot*, 146–147.

7. Ladas, Whipple, and Perry, *The G Spot*, 43.

8. Ladas, Whipple, and Perry, *The G Spot*, 60.

9. Ladas, Whipple, and Perry, *The G Spot*, 104, 106.

10. Martin Weisberg, "A Note on Female Ejaculation," *Journal of Sex Research* 17, no. 1 (February 1981): 90–91.

11. Ladas, Whipple, and Perry, *The G Spot*, 152.

12. Quoted in Irvine, *Disorders of Desire*, 165.

13. Ladas, Whipple, and Perry, *The G Spot*, 84.

14. Sabine zur Nieden, *Weibliche Ejakulation: Variationen zu einem uralten Streit der Geschlechter* (Gießen: Psychosozial-Verlag, 2009), 107.

15. Cf. Reimut Reiche, introduction to *Drei Abhandlungen zur Sexualtheorie*, by Sigmund Freud (Frankfurt: Fischer Verlag, 1991), 23.

16. Sigmund Freud, *Three Contributions to the Theory of Sex*, trans. A. A. Brill, 2nd ed. (New York: Nervous and Mental Disease Publishing Co., 1920), 81.

17. Rebecca Chalker, *The Clitoral Truth: The Secret World at Your Fingertips* (New York: Seven Stories Press, 2000), 83.

18. Cf. Anne Koedt, *The Myth of the Vaginal Orgasm* (Boston: New England Free Press, 1970).

19. Alfred C. Kinsey, *Sexual Behavior in the Human Female* (Philadelphia: W. B. Saunders Company, 1953), 579–580, 592.

20. William H. Masters and Virginia E. Johnson, *Human Sexual Response* (Boston: Little, Brown and Company, 1966), 253.

21. Shere Hite, *The Hite Report* (New York: Seven Stories Press, 1976), 215.

22. Alice Schwarzer, *Der "kleine Unterschied" und seine großen Folgen: Frauen über sich—Beginn einer Befreiung* (Frankfurt: Fischer Verlag, 1975), 10.

23. Koedt, *The Myth of the Vaginal Orgasm.*

24. Simone de Beauvoir, *The Second Sex*, trans. Constance Borde and Sheila Malovany-Chevallier (New York: Vintage Books, 2009), 384.

25. Schwarzer, *Der "kleine Unterschied,"* 202, 195, 206–207.

26. Carla Lonzi, *Taci, anzi parla: Diario di una femminista* (Shut up. Or rather, speak: Diary of a feminist) (Milan: Scritti di Rivolta Femminile, 1978).

27. Germaine Greer, *The Female Eunuch* (New York: McGraw-Hill, 1970), 31–32. Another paean to the vulva and vagina can be found in Greer's essay "Lady Love Your Cunt," in Germaine Greer, *The Madwoman's Underclothes: Essays and Occasional Writings* (New York: Atlantic Monthly Press, 1986). "Discovery that all female orgasm originates in the clitoris has been treated as discovery that the cunt is irrelevant. Not so" (76).

28. Ladas, Whipple, and Perry, *The G Spot*, 149.

29. Quoted in Shannon Bell, "Feminist Ejaculations," in *The Hysterical Male: New Feminist Theory*, ed. Arthur Kroker and Marielouise Kroker (New York: St. Martin's Press, 1991), 163.

30. In her classic polemic against pornography, Andrea Dworkin writes that "in such a symptomatic detail as ejaculating sperm," the Marquis de Sade's "libertine women are men." Female ejaculation is inconceivable to Dworkin. See Andrea Dworkin, *Pornography: Men Possessing Women* (New York: Plume, 1979), 95.

31. Sabine zur Nieden, "Die potente Frau," *EMMA* 10 (October 1987): 50–55.

32. Bea Trampenau, "Der Freudenfluß—die dritte Dimension des Orgasmus?," interview by Rita Götze, *clio* 41 (1995): 24.

33. Desmond Heath, "An Investigation into the Origins of a Copious Vaginal Discharge during Intercourse: 'Enough to Wet the Bed'—That 'Is Not Urine,'" *Journal of Sex Research* 20, no. 2 (May 1984): 194.

34. Cynthia Jayne, "Freud, Grafenberg, and the Neglected Vagina: Thoughts Concerning an Historical Omission in Sexology," *Journal of Sex Research* 20, no. 2 (May 1984): 212–215.

35. Heath, "An Investigation," 194, 196.

36. Cf. Frank Addiego, Edwin G. Belzer, Jill Comolli, William Moger, John D. Perry, and Beverly Whipple, "Female Ejaculation: A Case Study," *Journal of Sex Research* 17, no. 1 (February 1981): 13–21; Edwin G. Belzer, "Orgasmic Expulsions of Women: A Review and Heuristic Inquiry," *Journal of Sex Research* 17, no. 1 (February 1981): 1–12.

37. Joseph G. Bohlen, "'Female Ejaculation' and Urinary Stress Incontinence," *Journal of Sex Research* 18, no. 4 (November 1982): 360–363.

38. Edwin G. Belzer, Beverly Whipple, and William Moger, "On Female Ejaculation," *Journal of Sex Research* 20, no. 4 (November 1984): 403–406.

39. John D. Perry and Beverly Whipple, "Pelvic Muscle Strength of Female Ejaculators: Evidence," *Journal of Sex Research* 17, no. 1 (February 1981): 22.

40. Heli Alzate and Zwi Hoch, "The 'G Spot' and 'Female Ejaculation': A Current Appraisal," *Journal of Sex and Marital Therapy* 12, no. 3 (Fall 1986): 211.

41. Desmond Heath, "Female Ejaculation: Its Relationship to Disturbances of Erotic Function," *Medical Hypotheses* 24, no. 1 (September 1987): 103–106.

42. Marc A. Winton, "Editorial: The Social Construction of the G Spot and Female Ejaculation," *Journal of Sex Education and Therapy* 15, no. 3 (1989): 158.

43. Stifter tests and compares urine and ejaculate samples from forty men and women. Five women emit fluid right before orgasm that differs significantly from the urine of all forty participants, displaying distinct levels of prostatic acid phosphatase, urea, creatinine, and even glucose. Stifter is also the first to determine glutamate oxaloacetate transaminase and glutamate pyruvate transaminase values in ejaculate. The orgasmic fluids are "indisputably glandular secretions." Karl Stifter, *Die dritte Dimension der Lust: Das Geheimnis der weiblichen Ejakulation* (Munich: Heyne Verlag, 1988), 174–176. Translated by Elisabeth Lauffer.

44. Stifter, *Die dritte Dimension der Lust*, 185.

45. Carol Anderson Darling, J. Kenneth Davidson, and Colleen Conway-Welch, "Female Ejaculation: Perceived Origins, the Grafenberg Spot/Area, and Sexual Responsiveness," *Archives of Sexual Behavior* 19, no. 1 (February 1990): 29–47. The test subjects were "professional women in health-related fields."

46. Francisco Cabello Santamaría, "Female Ejaculation, Myth and Reality," *Sexuality and Human Rights: Proceedings of the XIIIth World Congress of Sexology* (Valencia, 1997): 326.

47. Alberto Rubio-Casillas and Emmanuelle A. Jannini, "New Insights from One Case of Female Ejaculation," *Journal of Sexual Medicine* 8, no. 12 (December 2011): 3503.

48. In their review of articles on various fluid expulsion phenomena in women during sex, the authors conclude that some wet orgasms do involve coital incontinence, in addition to ejaculation and squirting. Cf. Zlatko Pastor and Roman Chmel, "Differential Diagnostics of Female 'Sexual' Fluids: A Narrative Review," *International Urogynecological Journal* 29, no. 4 (December 2017): 621–629.

49. Felix D. Rodriguez, Amarilis Camacho, Stephen J. Bordes, Brady Gardner, Roy J. Levin, and R. Shane Tubbs, "Female Ejaculation: An Update on Anatomy, History, and Controversies," *Clinical Anatomy* 34, no. 1 (January 2021): 103, 106.

50. Ryoei Hara, Atsushi Nagai, Tohta Nakatsuka, Shin Ohira, Tomohiro Fujii, and Yosiyuki Miyaji, "Male Squirting: Analysis of

One Case Using Color Doppler Ultrasonography," *IJU Case Reports* 1, no. 1 (October 2018).

51. Josephine Lowndes Sevely, *Eve's Secrets: A New Theory of Female Sexuality* (New York: Random House, 1987), 39.

52. Sevely, *Eve's Secrets*, 138, 132, 120, 36.

53. "If it is recognized as such, the similarity between the woman's glans and the man's becomes obvious. Both have an overall acorn shape; both are perforated by the urethral opening." Sevely, *Eve's Secrets*, 22.

54. "In the female, the prostatic glands are dispersed along the floor of the urethra. Their distribution lengthwise is highly variable—a variability that has been attributed to a number of possible factors, including parity (the fact of having borne offspring), hormonal makeup, and age. In adult women, these structures are found in greater abundance toward the bladder end of the urethra; in the newborn, they are more likely to exist at the opposite end, closer to the urethral opening. Whatever the age of the female, however, these glandular structures are consistently located on the floor of the urethra, thus creating a bulge that pushes into the vagina." Sevely, *Eve's Secrets*, 45–46.

55. Sevely, *Eve's Secrets*, 47, 113.

56. Sevely, *Eve's Secrets*, 142. Sevely discusses male and female orgasm without ejaculation.

57. Sevely, *Eve's Secrets*, 91, 92, 93, 95, 180–181.

58. Sevely, *Eve's Secrets*, 90, 103. For more on vaginal innervation, see Per Olov Lundberg, "Die periphere Innervation der weiblichen Genitalorgane," *Sexuologie* 9 (2002): 99–106.

59. Bini Adamczak, "Come on. Über ein neues Wort, das sich aufdrängt—und unser Sprechen über Sex revolutionieren wird," *ak—analyse und kritik—Zeitung für linke Debatte und Praxis* 614 (March 15, 2016).

60. Sevely, *Eve's Secrets*, xxiii.

61. Terence M. Hines, "The G-Spot: A Modern Gynecologic Myth," *American Journal of Obstetrics and Gynecology* 185, no. 2 (August 2001): 359–362.

62. Emmanuele A. Jannini, Beverly Whipple, Sheryl A. Kingsberg, Odile Buisson, Pierre Foldès, and Yoram Vardi, "Who's Afraid of the G-Spot?," *Journal of Sexual Medicine* 7, no. 1 (January 2010): 29.

63. Cf. Amichai Kilchevsky, Yoram Vardi, Lior Lowenstein, and Ilan Gruenwald, "Is the Female G Spot Truly a Distinct Anatomic Entity?," *Journal of Sexual Medicine* 9, no. 3 (March 2012): 1–8. In 2014, meanwhile, another team dissects eight female cadavers and discovers the erotogenic zone in all of them. They figure the G-spot must exist—a conclusion others still reject. See A. Ostrzenski, P. Krajewski, P. Ganjei-Azar, A. J. Wasiutynski, M. N. Scheinberg, S. Tarka, and M. Fudalej, "Verification of the Anatomy and Newly Discovered Histology of the G Spot Complex," *An International Journal of Obstetrics and Gynaecology* 121, no. 11 (October 2014): 1333–1339.

64. "G-SHOT® by Dr. Matlock," accessed September 12, 2023, https://www.drmatlock.com/in-office-procedures-beverly-hills/g-shot.

65. Ada Borkenhagen and Heribert Kentenich, "Intimchirurgie: Ein gefährlicher Trend," *Deutsches Ärzteblatt* 11 (2009): A 502.

66. Beverly Whipple, "G Spot," in *The International Encyclopedia of Human Sexuality*, ed. Patricia Whelehan and Anne Bolin (Chichester, UK: Wiley-Blackwell, 2015), 1:429, 428.

67. "FIPAT—About," International Federation of Associations of Anatomists, last modified January 2020, https://ifaa.net/committees/anatomical-terminology-fipat/fipat-about/.

68. Milan S. Zaviačič, T. Zaviačič, R. J. Ablin, J. Breza, and K. Holoman, "The Female Prostate: History, Functional Morphology and Sexology Implications," *Sexologies* 11, no. 41 (2001): 44.

69. Zaviačič et al., "The Female Prostate."

70. Florian Wimpissinger, "Die weibliche Prostata—Faktum oder Mythos," *Urologie* 2, no. 7 (2007): 19.

71. Katarina S. Richterova, "Milan Zaviacic: The Slovak Scientist Who Discovered the Female Prostate," Radio Prague International, January 27, 2006, https://english.radio.cz/milan-zaviacic-slovak-scientist-who-discovered-female-prostate-8622301; Milan S. Zaviačič, "Die weibliche Prostata: Orthologie, Pathologie, Sexuologie und forensisch-sexuologische Implikationen," *Sexuologie* 9 (2002): 107–115.

72. Milan Zaviačič and R. J. Ablin, "The Female Prostate and Prostate-Specific Antigen: Immunohistochemical Localization, Implications of This Prostate Marker in Women and Reasons for Using the Term 'Prostate' in the Human Female," *Histology and Histopathology* 15, no. 1 (2000): 131.

73. By comparison, the male prostate weighs four to five times as much. The female urethra measures between approximately 1 and 1.5 inches.

74. In addition to the anterior or meatal type, Zaviačič identifies the posterior type, duct complex running the length of the urethra, rudimentary ducts, duct complex around the middle of the urethra, and a dumbbell-shaped configuration.

75. Zaviačič et al., "The Female Prostate," 48.

76. Zaviačič, "Die weibliche Prostata: Orthologie, Pathologie, Sexuologie," 109.

77. Zaviačič, "Die weibliche Prostata: Orthologie, Pathologie, Sexuologie," 111.

78. Wimpissinger, "Die weibliche Prostata—Faktum oder Mythos," 18.

79. Florian Wimpissinger, Karl Stifter, Wolfgang Grin, and Walter Stackl, "The Female Prostate Revisited: Perineal Ultrasound and Biochemical Studies of Female Ejaculate," *Journal of Sexual Medicine* 4, no. 5 (July 2007): 1391.

80. Florian Wimpissinger, Robert Tscherney, and Walter Stackl, "Magnetic Resonance Imaging of Female Prostate Pathology," *Journal of Sexual Medicine* 6, no. 6 (June 2009): 1704–1711.

81. Wimpissinger et al., "The Female Prostate Revisited."

82. Wimpissinger, "Die weibliche Prostata—Faktum oder Mythos," 20.

83. Wolf Dietrich, Martin Susani, Lukas Stifter, and Andrea Haitel, "The Human Female Prostate—Immunohistochemical Study with Prostate-Specific Antigen, Prostate-Specific Alkaline Phosphatase, and Androgen Receptor and 3-D Remodeling," *Journal of Sexual Science* 8, no. 10 (August 2011): 2818.

84. FIPAT, *Terminologia Anatomica*, 2nd ed. (Federative International Programme for Anatomical Terminology, 2019), fipat.library .dal.ca/TA2/.

85. For a thorough criticism of O'Connell's findings, see Vincenzo Puppo, "Anatomy of the Clitoris: Revision and Clarifications about the Anatomical Terms for the Clitoris Proposed (without Scientific Bases) by Helen O'Connell, Emmanuele Jannini, and Odile Buisson," *ISRN Obstetrics and Gynecology* 2011 (January 2011): 1–5.

86. Lisa Jean Moore and Adele E. Clarke, "Clitoral Conventions and Transgressions: Graphic Representations in Anatomy Texts, 1900–1991," *Feminist Studies* 21, no. 2 (Summer 1995): 284. Moore and Clarke examine the representation of the clitoris in anatomy books published between 1900 and 1991. These medical guides, they discover, depict and describe the clitoris in incomplete terms—if at all. The authors conclude that the field of anatomy has ignored feminist and other research findings that present new interpretations of the clitoris.

87. In Germany, for instance, *DER SPIEGEL* dedicates a page to O'Connell's findings ("Empfindsame Zwiebel," *DER SPIEGEL*, October 8, 1998). The European public service network ARTE broadcasts a documentary film on her work several years later (*Klitoris—Die schöne Unbekannte*, created by Michèle Dominici, directed by Stefan Firmin and Variety Moszynsky, first aired January 1, 2004).

88. According to O'Connell, the clitoris measures up to 3.5 inches long and 2.5 inches wide. The visible crown (*Glans clitoridis*, with its shaft and foreskin) continues inside the body in a pyramidal mass 1 to 1.5 inches long (*Corpus cavernosum*). This extends into a pair of crura

(*Crus clitoridis*) that form a V pointing toward the thighs. Two erectile bodies (*Bulbus vestibuli*) connect the pyramid and crura, which are themselves composed of spongelike tissue that swells with blood during sexual arousal. Individual erectile features are connected to the urethra, vagina, and one another by blood vessels and nerves.

89. Helen O'Connell, Kalavampara V. Sanjeevan, and John M. Hutson, "Anatomy of the Clitoris," *Journal of Urology* 174, no. 4 (October 2005): 1189.

90. Cf. Sharon Mascall, "Time for Rethink on the Clitoris," *BBC News*, last modified June 11, 2006, http://news.bbc.co.uk/2/hi/health/5013866.stm.

91. Helen O'Connell, Norm Eizenberg, Marzia Rahman, and Joan Cleeve, "The Anatomy of the Distal Vagina: Towards Unity," *Journal of Sexual Medicine* 8, vol. 5 (August 2008): 1886, 1883.

92. Nathan Hoag, Janet R. Keast, and Helen O'Connell, "The 'G Spot' Is Not a Structure Evident on Macroscopic Anatomic Dissection of the Vaginal Wall," *Journal of Sexual Medicine* 14, no. 12 (December 2017): 1532. In an encyclopedia article, however, O'Connell writes, "The anterior vaginal wall contains a sexually sensitive area, termed the Grafenberg spot, or G spot." See Helen O'Connell, "Anatomy, Female," in *The International Encyclopedia of Human Sexuality*, ed. Patricia Whelehan and Anne Bolin (Chichester, UK: Wiley-Blackwell, 2015): 1:74.

93. "The paraurethral glands (also termed lesser vestibular glands) are tubular glands that open into the distal urethra, and/or on each side of the urethral meatus." O'Connell, "Anatomy, Female," 1:72.

94. While working on her doctoral dissertation in Rwanda, Swedish researcher Jessica Påfs becomes fascinated by the everyday role kunyaza plays in sexuality there, whereas ejaculation and squirting are barely known or studied in her home country. In a 2021 study, Påfs surveys twenty-eight Swedish women about their experiences with coming. She's convinced that "the ignorance, and myths, surrounding the subject, is [*sic*] yet an aspect of oppressing women's sexuality," and she hopes "to bring awareness that this is a part of sexual response." Jessica Påfs, email to author, November 9, 2021. Cf. Jes-

sica Påfs, "A Sexual Superpower or a Shame? Women's Diverging Experiences of Squirting/Female Ejaculation in Sweden," *Sexualities* (September 2021).

95. Nsekuye Bizimana, "Another Way for Lovemaking in Africa: *Kunyaza*, a Traditional Sexual Technique for Triggering Female Orgasm at Heterosexual Encounters," *Sexologies* 19, no. 3 (June–September 2010): 162.

96. Barbara Achermann, *Frauenwunderland: Die Erfolgsgeschichte von Ruanda* (Ditzingen: Reclam, 2018), 148.

97. Nsekuye Bizimana, *Kunyaza: Multiple Orgasmen und weibliche Ejakulation mit afrikanischer Liebeskunst* (Freiburg: Hans-Nietsch-Verlag, 2009), 43, 44–45.

CHAPTER 6

1. Shannon Bell, *Whore Carnival* (New York: Autonomedia, 1995), 52.

2. Reyhan Şahin, a Turkish German rapper who goes by Lady Bitch Ray, writes that the more explicitly women discuss sex, the more space they demand for themselves. The same goes for those who ejaculate: "To many it's like metaphorical squirting when women take up space with their punani, when they talk about their vulva, vagina, pussy, or cunt on their own terms, or even pornographically. . . . For a lot of people it's like a facial cumshot." See Reyhan Şahin, "Sex," in *Eure Heimat ist unser Albtraum*, ed. Fatma Aydemir and Hengameh Yaghoobifarah (Berlin: Ullstein Verlag, 2019), 156. Translated by Elisabeth Lauffer.

3. Shannon Bell, *Fast Feminism* (New York: Autonomedia, 2010), 32; Bebe O'Shea, "The Dirt on the Squirt," *TORO magazine* (March 2006): n.p.

4. Rebecca Chalker, *The Clitoral Truth: The Secret World at Your Fingertips* (New York: Seven Stories Press, 2000), 110.

5. Bell, *Whore Carnival*, 263.

6. Bell, *Fast Feminism*, 38.

7. Bell, *Fast Feminism*, 39.

8. Bell, *Whore Carnival*, 264.

9. Annie Sprinkle, foreword to *Female Ejaculation and the G-Spot*, by Deborah Sundahl (Alameda, CA: Hunter House, 2003), ix.

10. Bell, *Fast Feminism*, 43.

11. Shannon Bell, "Feminist Ejaculations," in *The Hysterical Male: New Feminist Theory*, ed. Arthur Kroker and Marielouise Kroker (New York: St. Martin's Press, 1991), 35.

12. Bell, *Fast Feminism*, 35.

13. Bell, *Fast Feminism*, 40, 41.

14. Bell, *Fast Feminism*, 48.

15. "These days, when women ask me if it's worth learning to ejaculate, I answer that it is. But expect to do extra loads of laundry." See Annie Sprinkle, "The G Spot," Annie Sprinkle, accessed February 22, 2018, https://anniesprinkle.org/the-g-spot/.

16. Andrea Juno and V. Vale, *Angry Women* (San Francisco: RE/Search Publications, 1991), 23, 27.

17. Annie Sprinkle, *Hardcore from the Heart: The Pleasures, Profits and Politics of Sex in Performance* (London: Continuum, 2001), 49, 52.

18. Sprinkle, *Hardcore*, 74.

19. Linda Williams, *Hard Core: Power, Pleasure, and the "Frenzy of the Visible"* (Berkeley: University of California Press, 1989), 268.

20. Sprinkle, foreword to *Female Ejaculation*, ix.

21. Sprinkle, *Hardcore*, 8.

22. Annie Sprinkle, Veronica Vera, Frank Moores, Candida Royale, and Leigh Gates, "The Post Porn Modernist Manifesto," in *Film Manifestos and Global Cinema Cultures: A Critical Anthology*, ed. Scott MacKenzie (Berkeley: University of California Press, 2014), 382.

23. Mithu M. Sanyal, *Vulva: Die Enthüllung des unsichtbaren Geschlechts* (Berlin: Verlag Klaus Wagenbach, 2009), 181.

24. Emma L. E. Rees, *The Vagina: A Literary and Cultural History* (New York: Bloomsbury Academic, 2013), 281.

25. Annie Sprinkle, "A Public Cervix Anouncement [*sic*]," Annie Sprinkle, accessed April 21, 2018, http://anniesprinkle.org/a-public-cervix-anouncement/.

26. Sprinkle, *Hardcore*, 93.

27. Cf. Sprinkle, "The G Spot."

28. Cf. Federation of Feminist Women's Health Centers, *A New View of a Woman's Body: A Fully Illustrated Guide* (New York: Simon and Schuster, 1981), 54.

29. Cf. Sprinkle, "The G Spot."

30. Sprinkle, foreword to *Female Ejaculation*, xi, xii, xi.

31. Chris Straayer, "The Seduction of Boundaries: Feminist Fluidity in Annie Sprinkle's Art/Education/Sex," in *More Dirty Looks: Gender, Pornography and Power*, ed. Pamela Church Gibson (London: British Film Institute, 2004), 234–235.

32. Quoted in Straayer, "The Seduction of Boundaries," 234.

33. Sprinkle, foreword to *Female Ejaculation*, xii.

34. Annie Sprinkle and Beth Stephens, "Ecosex Manifesto," Sprinkle and Stephens Collaboration, accessed February 24, 2022, https://sprinklestephens.ucsc.edu/research-writing/ecosex-manifesto/.

35. Annie Sprinkle, Beth Stephens, and Jennie Klein, *Assuming the Ecosexual Position: The Earth as Lover* (Minneapolis: University of Minnesota Press, 2021).

36. Sprinkle and Stephens, "Ecosex Manifesto."

37. Deborah Sundahl, *Female Ejaculation and the G-Spot* (Alameda, CA: Hunter House, 2003).

38. Quoted in Jill Nagle, ed., *Whores and Other Feminists* (New York: Routledge, 2013), 158.

39. Kayla Ginsburg, *On Our Backs with a Bad Attitude*, video, 8:16 min., accessed February 23, 2018, https://www.youtube.com/watch?v=XStx7V1n79E&ab_channel=KaylaGinsburg.

40. In this fifteen-minute video, Lane, a San Francisco–based sex educator and consultant, tells the history of female ejaculation, explains female sexual anatomy, talks about her own experiences with ejaculation, masturbates, and comes on screen.

41. Sundahl, *Female Ejaculation*, 3, 105, 21.

42. "The G-spot is defined as both the prostate and a network of erectile tissue [that] extends . . . beyond the G-spot. . . . Therefore, the G-spot is not merely a 'spot' on the wall of the vagina. Rather, it is an organ one can feel and stimulate *through* the vaginal wall." Sundahl, *Female Ejaculation*, 35–36.

43. Deborah Sundahl, "Das gynäkologische Rätsel," interview by Theresa Bäuerlein, *Krautreporter*, March 13, 2015, https://krautreporter.de/433-das-gynakologische-ratsel.

44. Quoted in Stephanie Osmanski, "Everything You Need to Know about Female Ejaculation, Straight from a Sex Expert," *Hello-flo*, June 26, 2017, https://helloflo.com/everything-you-need-to-know-about-female-ejaculation-straight-from-a-sex-expert/.

45. Sundahl, *Female Ejaculation*, 46.

46. Beeke, "Zwölf verschiedene Arten zum Orgasmus zu kommen," *Femna Health*, accessed July 31, 2019, femna.de/zwoelf-verschiedene-arten-zum-orgasmus-zu-kommen.

47. Sundahl, *Female Ejaculation*, xviii.

48. Quoted in Hugh B. Urban, *Magia Sexualis: Sex, Magic, and Liberation in Modern Western Esotericism* (Berkeley: University of California Press, 2006), 105.

49. David Gordon White, *Kiss of the Yoginī: "Tantric Sex" in Its South Asian Contexts* (Chicago: University of Chicago Press, 2006), 13.

50. Rufus Camphausen, *The Yoni: Sacred Symbol of Female Creative Power* (Rochester, VT: Inner Traditions, 1996).

51. In their book, the Muirs write that ejaculate is produced in one of the Bartholin's glands and expressed from the urethral opening. This claim is false.

52. Caroline Muir and Charles Muir, *Tantra: The Art of Conscious Loving* (San Francisco: Mercury House, 1989).

53. Sundahl, *Female Ejaculation*, 155, 180.

54. Elmar Zadra and Michaela Zadra, *Hingabe und Ekstase: Der G-Punkt und das Geheimnis der weiblichen Sexualität* (Munich: Droemer Knaur, 2004), 89, 38, 233, 104, 158.

55. Zadra, *Hingabe und Ekstase*, 119.

56. Zadra, *Hingabe und Ekstase*, 147.

57. "Stellungnahme des Tantramassage-Verbandes e.V. zum Prostituiertenschutzgesetz," Tantramassage-Verband e.V., last modified July 2, 2017, https://www.tantramassage-verband.de/wp-content/uploads/2017/07/Stellungnahme-TMV-Langversion-1.pdf, 8, 7.

58. Florian Wimpissinger, Christopher Springer, and Walter Stackl, "International Online Survey: Female Ejaculation Has a Positive Impact on Women's and Their Partners' Sexual Lives," *BJU International* 20 (2013): E180.

59. Osmanski, "Everything You Need to Know about Female Ejaculation."

60. Quoted in Urban, *Magia Sexualis*, 108.

61. Straayer, "The Seduction of Boundaries," 235.

62. The film *Female Ejaculation & Other Mysteries of the Universe* (2020) from the German filmmaker Julia Ostertag falls within this tradition as well. The independent production takes viewers on an autobiographical voyage into the world of female coming with such new and familiar faces as Sprinkle, Bell, Laura Méritt, Diana J. Torres, and Fluida Wolf. Alice Heit's 2019 film *Les eaux profondes* (*Deep Waters*), meanwhile, is a poetic examination of female coming. Mixing Super 8 footage with animation, the French filmmaker showcases a number of *femmes fontaines*.

63. "Squirting Searches," Pornhub, last modified November 11, 2017, https://www.pornhub.com/insights/squirting-searches.

64. Naomi Wolf, *Vagina: A New Biography* (New York: Ecco, 2012), 180.

65. Quoted in Marco Wedig, "Weibliche Ejakulation: Die Prostata ist für alle da," *die tageszeitung*, last modified April 25, 2015, https://www.taz.de/!5010876/.

66. Cited in Matthew D. LaPlante, "Cytherea's Comeback," *Las Vegas City Life*, last modified October 6, 2014, https://web.archive.org/web/20141006034935/lasvegascitylife.com/sections/news/cytherea%E2%80%99s-comeback-rise-and-fall-mormon-girl-who-charted-new-course-adult?i.

67. Williams, *Hard Core*, 48.

68. Astrid Dreßler, *Dildo, Peitsche, Latexhandschuh: Eine Filmanalyse lesbisch/queerer Pornografie* (Marburg: Tectum Verlag, 2015), 111.

69. In its digital lexicon of film terms, the University of Kiel applies the term "cum shot" to both male and female ejaculation. See *Das Lexikon der Filmbegriffe*, s.v. "cum shot (*n*.)," accessed March 3, 2018, https://filmlexikon.uni-kiel.de/doku.php/c:cumshot-6950.

70. Wikipedia, s.v. "Female Ejaculation," accessed March 1, 2022, https://en.wikipedia.org/wiki/Female_ejaculation.

71. Quoted in LaPlante, "Cytherea's Comeback."

72. Quoted in EJ Dickson, "Cytherea the Squirt Queen Is Making Her Return to Porn," *Daily Dot*, last modified April 17, 2020, https://www.dailydot.com/irl/cytherea-the-squirt-queen/.

73. Quoted in Dian Hanson, *The Big Book of Pussy* (London: Taschen, 2011), 280.

74. Quoted in Hanson, *The Big Book of Pussy*, 283, 287.

75. Rebecca Saunders, "Open Wide and Say Aaahh! Female Ejaculation in Contemporary Pornography," in *The Sexualized Body and the Medical Authority of Pornography: Performing Sexual Liberation*, ed. Heather Brunskell-Evans (Cambridge, UK: Cambridge Scholars Publishing, 2016), 99.

76. Sundahl, "Das gynäkologische Rätsel."

77. Cf. Straayer, "The Seduction of Boundaries"; Sharon Moalem, "Everything You Always Wanted to Know about Female Ejaculation (but Were Afraid to Ask)," *New Scientist*, last modified May 27, 2009,

https://www.newscientist.com/article/mg20227101-200-every thing-you-always-wanted-to-know-about-female-ejaculation-but -were-afraid-to-ask/.

78. Bella Counihan, "Weird Politics of Small Boobs and Bodily Fluids," *Sydney Morning Herald*, January 29, 2010, https://www .smh.com.au/politics/federal/weird-politics-of-small-boobs-and -bodily-fluids-20100129-n278.html.

79. Kristina Lloyd, "Sexuality, as Defined by Censors," *Guardian*, last modified October 8, 2009, https://www.theguardian.com /commentisfree/2009/oct/08/pornography-sexuality-censors -female-ejaculation.

80. Deborah Sundahl, "On the 2014 Ruling by the UK Censorship Board on Female Ejaculation," *Deborah Sundahl* (blog), last modified March 13, 2015, https://deborahsundahl.com/on-the-2014-ruling -by-the-uk-censorship-board-on-female-ejaculation/.

81. Hanna Rosin, "The 'Myth' of Female Ejaculation," *Slate*, last modified June 4, 2009, https://slate.com/human-interest/2009/06 /the-myth-of-female-ejaculation.html. Emphasis in original.

82. "Female Director Victorious with First Ever UK Release of a Film That Contains Female Ejaculation," *International Entertainment News*, last modified October 6, 2009, https://www.prnewswire .co.uk/news-releases/female-director-victorious-with-first-ever-uk -release-of-a-film-that-contains-female-ejaculation-152576265.html.

83. Lloyd, "Sexuality, as Defined by Censors."

84. "Female Director Victorious."

85. Quoted in Hanson, *The Big Book of Pussy*. Bukkake is a group sex act often depicted in porn, in which several men ejaculate on another person, usually a woman.

EPILOGUE

1. "Pussy-Profile" (PDF), Million Puzzies, accessed May 22, 2019, http://www.millionpussyproject.com/en/index.php#.

2. For more information and upcoming dates, see https://weib lichequelle.de/en/.

3. Laura Méritt, conversation with author, April 26, 2019.

4. Méritt, conversation with author.

5. Cf. Catherine Blackledge, *The Story of V: Opening Pandora's Box* (London: Weidenfeld and Nicholson, 2003), 297.

6. Daniel Bergner, *Die versteckte Lust der Frauen: Ein Forschungsbericht* (Munich: Knaus Verlag, 2014), 238.

7. Cf. Sharon Moalem and Joy S. Reidenberg, "Does Female Ejaculation Serve an Antimicrobial Purpose?," *Medical Hypotheses* 73, no. 6 (September 2009): 1069–1071.

8. Sabine zur Nieden, *Weibliche Ejakulation: Variationen zu einem uralten Streit der Geschlechter* (Gießen: Psychosozial-Verlag, 2009), 35–36.

9. Barbara Ehret and Mirjam Roepke-Buncsak, *Frauen Körper Gesundheit Leben: Das große BRIGITTE-Buch der Frauenheilkunde* (Munich: Diana Verlag, 2008), 89.

10. In summer 2021, Berlin-based artist Julia Frankenberg takes the female prostate to the streets in Popsicle form. In flavors like curaçao-lime–passion fruit or oat milk–black currant, her frozen treats come in many different (prostatic) shapes. Frankenberg accepts donations for the homemade "Squirt Eis," which she wheels around in a bicycle rickshaw and hands out along with educational materials on the prostate. It's edible art, a "mobile, lickable, ephemeral sculptural intervention." The project combines sex ed with summer fun. Frankenberg engages hundreds of people in conversation about the prostate and ejaculation in vulva-havers. See Squirt-Eis, accessed December 31, 2021, https://squirt-eis.org/.

11. The German government's Federal Center for Health Education, part of the Ministry of Health, hosts a teen-oriented sex ed site called Loveline. The definition provided for clitoris is incomplete and pretty vague: "'Clitoris' is the full medical term for 'clit.' The clitoris is a female sexual organ. The visible tip of the clitoris is located between the inner lips. That's just the part of the clitoris you can see—this organ can actually extend up to 4 inches and runs along the pubic bone. The clitoris is a very sensitive organ. For many women, the visible tip of the clitoris is the most important point of stimulation to achieve orgasm." The prostate and ejaculation are

defined solely with regard to boys/men. See "#LEXIKON–K," Loveline, accessed March 23, 2018, https://www.loveline.de/.

12. "Clitoris," Fab Lab, accessed December 12, 2021, http://carrefour-numerique.cite-sciences.fr/wiki/doku.php?id=projets:clitoris.

13. Daniel Haag-Wackernagel, email to author, August 8, 2020. Haag-Wackernagel stresses that "almost nothing is known about the bulbo-clitoral system, even in medical circles."

14. See Michael Schünke, Erik Schulte, Udo Schumacher, Markus Voll, Karl H. Wesker, and Nathan F. Johnson, eds., "Location, Structure, and Innervation of the Bulboclitoral Organ," *THIEME Atlas of Anatomy, Volume 1: General Anatomy and Musculoskeletal System*, 3rd ed. (New York: Thieme, 2020), https://medone-education.thieme.com/ebooks/cs_11302222?context=search/0#/ebook_cs_11302222_cs21360.

15. Jessica Pin (@jessica_ann_pin), "I Have AMAZING NEWS!," Instagram, photo, February 26, 2023, https://www.instagram.com/p/CpJwncMu6J8/.

16. Cf. Richard Drake, A. Wayne Vogl, and Adam W. M. Mitchell, *Gray's Anatomy for Students*, 3rd ed. (Philadelphia: Churchill Livingstone, 2015); Frank H. Netter, *Atlas der Anatomie* (Munich: Elsevier, 2011); Frank H. Netter, *Gynäkologie* (Stuttgart: Thieme, 2006); Michael Schünke, Erik Schulte, Udo Schumacher, Markus Voll, and Karl Wesker, *Prometheus: LernAtlas der Anatomie, Innere Organe* (Stuttgart: Thieme, 2012). One exception is Hans-Joachim Ahrendt and Cornelia Friedrich, eds., *Sexualmedizin in der Gynäkologie* (Heidelberg: Springer Verlag, 2015), 14: "Strictly speaking, female ejaculation is the emission of small amounts of whitish fluid from the female prostate, or Skene's paraurethral glands. This has been proven by the presence of prostatic acid phosphatase in female ejaculate. The 'female prostate' is an exocrine organ, the location and size of which varies, found in 2/3 of all women. It is typically located in the distal half of the vagina, to the side of the urethra. The significance of the female prostate is not yet entirely clear. It produces a milky secretion expressed primarily during vaginal and clitoral stimulation."

17. Petra Bentz, email to the author, May 3, 2019.

18. Landing page of Paulita Pappel's porn video *Female Ejaculation*, accessed May 16, 2018, xconfessions.com/film/female-ejaculation; Rebecca Saunders, "Open Wide and Say Aaahh! Female Ejaculation in Contemporary Pornography," in *The Sexualized Body and the Medical Authority of Pornography: Performing Sexual Liberation*, ed. Heather Brunskell-Evans (Cambridge, UK: Cambridge Scholars Publishing, 2016), 96.

19. Cf. Sandra R. Leiblum and Rachel Needle, "Female Ejaculation: Fact or Fiction," *Current Sexual Health Reports* 3, no. 2 (June 2006): 85–88; Emmanuele A. Jannini, Giulia d'Amati, and Andrea Lenzi, "Histology and Immunohistochemical Studies of Female Genital Tissue," in *Women's Sexual Function and Dysfunction: Study, Diagnosis and Treatment*, ed. Irvin Goldstein, Cindy M. Meston, Susan Davis, and Abdulmaged Traish (Philadelphia: Taylor and Francis, 2005), 125–133.

20. Florian Wimpissinger, Christopher Springer, and Walter Stackl, "International Online Survey: Female Ejaculation Has a Positive Impact on Women's and Their Partners' Sexual Lives," *BJU International* 20 (2013): E180.

21. Wimpissinger, Springer, and Stackl, "International Online Survey," E180; Nieden, *Weibliche Ejakulation*, 113.

22. For a wonderful analysis and appreciation of female genital smells, read "The Perfumed Garden," a chapter in Blackledge's *The Story of V*.

23. Laurie Penny, *Meat Market: Female Flesh under Capitalism* (Winchester, UK: Zero Books, 2011).

24. Nieden, *Weibliche Ejakulation*, 114.

25. Cf. Wiebke Bolle and Pia Seitler, "Periode abschaffen—Frauen erzählen, welche Erfahrungen sie gemacht haben," *DER SPIEGEL*, last modified March 29, 2019, https://www.spiegel.de/panorama/pille-durchnehmen-welche-folgen-hat-ein-langzyklus-zwei-frauen-und-eine-aerztin-erzaehlen-a-d78ea323-174e-4df9-a63f-0fbf6c8a6c2c.